Orchard Irrigation and Soil Management Practices

THE AUTHOR

Prof. Bibhas Chandra Mazumdar had been in the profession of teaching of post-graduate and under-graduate students of several Indian Universities for more than 34 years and supervised researches towards 72 theses of post-graduate students and doctoral fellows. He had also served in fruit development, extension education, farm management and in research positions for more than 7 years. Singly or jointly with others, he has authored 14 books, 163 research papers and 54 non-research articles in horticulture, human nutrition, medicinal plants and plant physiology.

Orchard Irrigation and Soil Management Practices

2nd Revised and Enlarged Edition

– Author –

Dr. Bibhas Chandra Mazumdar

M.Sc. Agri., Ph.D.
Former, Professor in Horticulture,
Institute of Agricultural Sciences, Calcutta University;
Reader in Horticulture, Bidhan Chandra Krishi Vishwavidyalaya,
Kalyani and Mohanpur, West Bengal;
Lecturer in Horticulture, Allahabad Agricultural Institute (Now SHUATS),
Allahabad, Uttar Pradesh;
Fruit Development Officer, Govt. of West Bengal.

2025

Daya Publishing House®
A Division of
Astral International Pvt. Ltd.
New Delhi–110002

ISBN: 978-93-5919-974-0

Published by : **Daya Publishing House®**
A Division of
Astral International Pvt. Ltd.
– ISO 9001:2015 Certified Company –
4736/23, Ansari Road, Darya Ganj
New Delhi-110 002
Ph. 011-43549197, 23278134
e-mail: info@astralint.com
Website: www.astralint.com

Dedicated to

The venerated memory of the Late (Dr.) John C. Edwards, former Professor and Head, Departments of Plant Pathology and Horticulture at the then Allahabad Agricultural Institute (now SHUATS, Allahabad, U.P.), a genius in various branches of plant science of the last century. His kind affection to the author would hardly be blurred.

Preface

The knowledge on the principles of fruit growing is the pre–requisite to gain expertise on the scientific husbandry of fruit crops. It is mentionable in this connection that although there have been a number of authentic books which have presented the procedural details of growing various types of fruit crops under Indian conditions, separate titles covering elaborate discussion pertaining to the principles and methods of fruit culture having appositeness to the country are scanty.

While considering the above gap, a title had been brought out by the author in the year 2004 incorporating two fundamental aspects in the area of principles of fruit culture, *i.e.*, orchard irrigation and soil management practices to meet the requirements of the post–graduate and advanced levels of the under–graduate students of horticulture in India keeping in view their curriculum of courses. Imparting practical knowledge to the fruit growers in India had also been an objective.

However, the science and technology involved in the orchard irrigation and the orchard floor management have greatly enriched in the recent years and many new avenues have been opened up in this area of pomology. Looking over these new additions, it has been attempted by the author to bring out the title in the present from making it revised and enlarged to a considerable extent.

The author is grateful to the students and research fellows with whom he had the opportunity to study pomology in his teaching and research career and the commercial fruit growers of West Bengal with whom he had interacted while he was a fruit development officer in this State, especially for the volleys of questions and also the practical suggestions of fruit growing put forward by them.

Sincere gratitude is due to Mrs. Purabi Roy Mazumdar, M. A., B. Ed. and Mrs. Kaushiki Mazumdar Dutta, M. Sc., Ph. D. for various assistance received in designing and shaping the present title.

The author welcomes valued suggestions from the teachers, researches and students of horticulture as well as those associated with fruit growing for further enrichment of the title in future.

Bibhas Chandra Mazumdar

Contents

Part II : ORCHARD SOIL MANAGEMENT PRACTICES

Part : I

ORCHARD IRRIGATION

Chapter 1

Necessity of Water to Human Beings

Water is the free gift of the Mother Earth. This invaluable material has its existene in liquid, solid and gaseous states and in pure form, it gives no taste and colour. In the past, water was chemically considered to be an element but it was later brought to light that the matter is an oxyhydride such that a molecule of it is composed of one oxygen atom which is covalently bonded to two hydrogen atoms.

The prime need for survival of human beings and all living organisms is water. In fact, there is adequate evidence that life had originated in oceans. No life can exist without water. In case of human beings, it is known that an individual can survive for upto 4 to 6 weeks without taking food but it is impossible to survive for more than 3 days without taking any form of water, which means that death will result if there is a loss of 20 per cent of body weight.

Human beings need water not only to sustain life but it is an indispensable item to them for progress, development, economy and civilization. Water cannot be avoided for quenching thirst, cooking of food, raising of crop plants and livestock management. Its necessity is felt to secure clothes, household commodities and building materials. Conservation of water is essential for pisciculture and obtaining aquatic foods. Forest cover and eco–environmental balance are not possible without water. These apart, water is an essential item to run a series of industries including hydro–electric, thermal and nuclear power generation.

Requirement of Water

Requirement of water by human beings varies according to countries, climate, agriculture, horticulture, industry, population, flora and fauna and other factors. Important countries which consume fairly large amount of water are China, India, USA and Japan.

In India, water consumption pattern is 76 per cent in agriculture and horticulture, 6.25 per cent in industries, 6.25 per cent in generation of power, 4.7 per cent in domestic use and for transport and other purposes, the consumption comes to 6.8 per cent.

Chapter 2

Water Content on Earth

Surface of the Earth remains to be covered with 70 per cent of water. However, of this amount, 97 per cent is sea or ocean water which in containing high concentration of salt is not considered fit for practical use. The remaining 3 per cent water is non–saline and is termed *fresh water.* But of this 3 per cent non–saline part, about 2 per cent is stored in the glaciers and ice–caps. Of the remaining 1 per cent, 0.7 per cent is ground water and 0.3 per cent constitute the water that is actually usable.

The 0.3 per cent usable water is stored in the rivers, streams, lakes, ponds, ditches, soils, drains, atmosphere and including the body of plants and animals. Under Indian condition, 90 per cent of the water is available from the rivers but it needs mentioning that all river waters in the country are polluted to a considerable extent.

How Earth Receives Water

The Earth receives water from the Earth itself and not from any extra–terrestrial object. Thus, it could be said to act like *energy* in that, the energy which was present at the time of birth of this planet remains the same even at this moment. In the same way, the water that is present in the whole of the Earth is only converted from one state to other and changes its position from one location to other. From this standpoint, water is regarded to be a *renewable natural source.* It functions in a *cyclic process*, which denotes that it goes and comes back and thereby, changes its form only.

The surface of the Earth receives water from *precipitation.*

Precipitation

The term, precipitation has been stated in the science of meteorology as a *condensation product of water vapour* of the atmosphere that drops on Earth surface from the clouds due to gravity. It occurs in different forms, *i.e.*, types, which are stated below.

Every year, about 505,000 cubic kilometres of water is deposited on Earth surface as precipitation. Out of this, about 398,000 cubic kilometres fall on seas, oceans, rivers, lakes, streams *etc.* and the rest 107,000 cubic kilometres are received by the Earth surface. That is, 21.2 per cent actually falls on Earth surface. It should be noted that precipitation is received not only by Earth but also Moon and some planets.

Forms of Precipitation

There are various forms or types of precipitation and the important ones are stated below.

(a) Rain

When the size of water droplets that fall on Earth surface from the clouds ranges from 1 millimetre to 9 millimetres, the discharge is referred to as rain. (According to some, the term rain should be restricted when the droplet size is round about 2 millimetre in diameter). When the size of rain droplets is increased, they assume oblate shape.

(b) Drizzle

Drizzle is rain of very low light intensity and the size of the droplets is less than 0.5 millimetre. For such small size, the drizzles evaporate before they reach the Earth surface.

(c) Mist and Fog

These are extremely fine droplets which are suspended in air. Fog may be considered as a cloud which lies at very low level. These disappear with sunlight.

(d) Rime

Rime occurs when fog of very low temperature strikes some object and forms white ice. It is stated to be freezing fog.

(e) Snow

Snow appears when water vapour sublimes at less than freezing temperature. These are ice–crystals which form while they are suspended in the atmosphere. They are modified after falling on ground. (A recent report states that rain has fallen on the top of Greenland's ice cap which is 2 miles above sea level. Temperature of the peak is much below freezing but rose above it for 9 hours and resulted rain for several hours during the middle of August in 2021).

(f) Sheet

Sheet is formed when rain freezes if it passes through a cold layer.

(g) Hail

This is a phenomenon of freezing of the rain drops which is formed due to severe condensation and joining, *i.e.*, coalescence of water droplets at much higher level of Earth surface. Among various types of precipitation, hail is the only solid form. Hail is considered different from ice–pellets (= sleets). It consists of lumps of ice and the each individual lump is stated as hailstone. Hails do not fall when surface temperature is low unlike ice–pellets which fall when temperature goes very low.

In India, total amount of precipitation is estimated to be 3,700,000 million cubic metres.

Rainfall Situation in India

Among different forms of precipitation, rainfall plays a great influence in the production of agricultural, horticultural and plantation crops in a country.

India's average annual rainfall is estimated to be about 1250 millimetre and this is somewhat greater than the world average of 990 mm. The country receives 75–80 per cent rainfall by south–west monsoon which ranges from the month of June to September and the rest 20–25 rainfall is received during the remaining eight months.

However, a number of constraints are faced by the country in regard to utilization of rain in growing crop plants.

Extreme spatial and temporal variation could be said to be a great handicap. It may be stated that while in the Khasi hills in the north–east India, average annual rainfall goes to 10,000 millimetres (Cherrapunji which is in that zone receives highest rainfall in the world where the average annual rainfall is round about 10,900 millimetres), parts of Rajasthan deserts, receive 150 millimetres annual rainfall.

Although rainfall appears to be satisfactory during four months from June to September due to south–west monsoon, it could be said, prolonged dry spells often occur in many parts of the country. Acid rain also occurs in some parts of the country in some years which results due to atmospheric pollutants, *e.g.*, nitrogen oxides, sulphur dioxide *etc.* It may be stated that normally, rain water is always slightly acidic (*p*H 5.5–6.5) due to formation of carbonic acid by the reaction of carbon di–oxide of atmosphere but if the acidity goes very low, the growth of the crop plants and their productiveness are hampered. Acid rains are common near industrial areas. Water–logging due to excessive rain occurs almost every year in some parts of India.

Chapter 3

Importance of Water in Plants, its Uptake and Conduction

We know that like human beings and animals, no plant can survive without water—whether the amount is small or large. The questions arise, what the plants do with the water, in other words, their physiological actions on receiving it, how they take up and send it to other parts. These aspects have been brought in this Chapter in brief.

Physiological Role

Among many physiological actions of water that take place in the plant body, some of the important ones may be put up as follows:

(i) Water is the ***fluid of life*** and brings about hydration of the protoplasm of the cells in the plant body.

(ii) In most of the higher group of plants, it comprises 90–95 per cent in leaves, 60–90 per cent in roots and 70–90 per cent in pulpy fruits.

(iii) It is essential for all *metabolic reactions.* Direct part of water is played in photolysis in the plant cells during the photosynthetic process.

(iv) For condensing and hydrolyzing reactions that are constantly accomplished within plant body, water is essential.

(v) Water acts as a *carrier* for uptake of oxygen and various nutrients in solution form by the plants from the soil, their translocation to different parts and thereafter, onward distribution of the synthesized products to the various organs in their body.

(vi) The *specific heat* of water is high and for this reason, rate of absorption or loss of heat by plant tissues is greater with little changes of temperature. The continuity in the process of absorption and dissipation of water by plants is of significant importance in protecting them from extremes of heat or cold.

(vii) The amount of synthesized (= anabolic) products when measured in terms of dry matter content of the fresh tissues are found to be controlled by the amount of water which is taken up by the plants. Different plants require different volumes of water to produce particular amount of dry matter in them and this is referred to as the *water requirement* of plant. This is specific to a particular plant and the environmental conditions to which it grows.

It is a common knowledge that the vascular plants which contain roots take up water as well as mineral matter mainly from the soil. Under certain conditions however, shoots also take up moisture from the atmosphere or absorb dew but these are very small and not of significance.

However, several phenomena are involved in the whole process of taking up water by a plant and its onward distribution in different parts of its body. These may be brought under the following categories.

(i) ***Absorption***, *i.e.*, uptake of water by the roots from the soil.

(ii) ***Upward translocation***, *i.e.*, conduction of the absorbed water in its body.

(iii) ***Downward translocation*** of the absorbed water and

(iv) ***Lateral translocation*** of the absorbed water. Knowledge of the above phenomena has its necessity for efficient water management of crop plants.

Absorption of Water

Water is absorbed, which means it is taken up by the roots of the plants that inhabit in the soil in higher plants. Hence, the subject may be discussed under the following categories.

(a) Types of Water in the Soil

Soil is a porous medium and water is held in it in the following four ways.

(i) ***Capillary water***: This is that form of soil water which is held (= fine retained) by the capillary pores (= fine pores) that are present in the soil. This water is actually *available* to the roots and it should be noted that almost all the water that takes up by the roots of a plant is capillary water.

(ii) ***Gravitational water***: This water is also known as *ground water*. Water of this form flows down freely due to gravity and flows to greater depth. At this depth, the roots of most plants and trees are in fact, unable to reach. Hence, gravitational water is not taken up by the roots.

(iii) ***Hygroscopic water:*** This water is very strongly held as thin films and hence, roots cannot draw up this water.

(iv) ***Chemically combined water:*** Water of this type remains in chemically combined form in the structure of the soil mineral matter and accordingly, this water is never absorbed by the plant roots.

(b) Water Absorbing Organs of Plants

Water is taken up by the terminal portions of the roots. However, maximum absorption of water takes place by the *root hairs*. These are located at 1–10 cm above the tip of each root. Structurally, a root hair is a tubular prolongation which arises at right angle to the axis of the piliferous cell of root. In the root hair zone, the xylem vessel is not fully mature and the epidermis and endodermis are permeable. The cell–wall in a root hair has two layers. The outer layer is rich in pectin and for this, the particles of the soil are firmly adhered (= attached) to it, while the inner layer of cell wall mainly consists of cellulose.

(c) Mechanism of Water Absorption by Roots

There are two types of water absorption, which are known as *active absorption* and *passive* absorption and they occur independently.

In active absorption, the force which is necessary to absorb water from soil originates directly in the *roots*. Hence, it is called active absorption. This absorption takes place in plants which *transpire poorly* or in plants which grow in *water* or in *marshy land*. On the other hand, in passive absorption, the force which is necessary to absorb water originates in the *shoots* and not in the roots and the force which originates in shoot is transmitted to the root hairs and due to that, absorption of water takes place.

Active absorption has been explained from osmotic theory and non–osmotic theory. Osmotic theory states that water of the soil is absorbed across a differentially permeable membrane. That is, as the osmotic pressure of soil water becomes lower than the osmotic pressure of cell sap, soil water is absorbed across the cytoplasmic membrane and which membrane is *semi–permeable* in nature. Thus, absoption of water could be said to be an *endo–osmosis* process.

In the other system, *i.e.*, non–osmotic absorption, it is the *energy* which causes absorption of water. The energy comes from respiration, *i.e.*, due to breakdown (= catabolism) of carbohydrate molecules and the energy that is liberated by that causes uptake of water into the xylem vessel. Passive absorption takes place in those plants which transpire vigorously.

In the cells of the leaves, the *diffusion pressure deficit* increases. Accordingly, water diffuses out from the xylem vessels and then it moves to the mesophyll cells of the leaves. Thereby, a tension is produced in the column of water that exists in the xylem. This tension continues upto the whole length and extends right upto the root hairs. In fact, large amount of water is absorbed by a land plant by passive absorption. It is observed that rate of transpiration is usually *proportional* to the absorption of water.

Upward Translocation of Water in Plants

When water is absorbed by the roots, it is sent, *i.e.*, translocated (= conducted) to the upper parts by the plants.

Although exact mechanism of upward translocation which is known as *ascent of sap* is yet to be precisely known, several theories have been put forward to explain the mechanism. These are stated as vital theories, root pressure theory and physical theory and these are presented below in brief.

(a) Vital Theories

These theories are based in consideration to the fact that although vessels are non–living but they are in close contact with cells which are in fact, living and vital activities of these living cells are responsible for ascent of sap. Several vital theories have been suggested by plant physiologists but these have shortcomings and hence, could not be much accepted, though cannot also be ignored at all.

Godlewski (1884) proposed the 'relay pump' theory. It has been suggested by him that a *rhythm–like change* had been apparent in the osmotic pressure of living cells of the xylem parenchyma and medullary rays. For this increase and decrease occurring repeatedly in the osmotic potential, water is forced, *i.e.*, pumped into the vessels and thus, water moves up in a *clambering manner*. This theory was not supported by Strasburger (1893) and many others as they observed that ascent of sap had not been dependent on living cells and this might take place even if the living cells could be destructed (= killed) by application of poisonous chemicals.

Sir J.C. Bose (1923) put forward a vital theory with an electric probe, a galvanometer and electric circuit. He put forward the *pulsation theory* and demonstrated that the living cells lying at the innermost layer of cortex at the outer side of the endodermis remained in some sort of pulsation in a rhythmic manner and due to this rhythmic *pulsation,* pumping of water occurs from one cell to the other such that water moves upward. This theory was supported by a section but discredited by Dixon and others. According to the latter scientists, translocation to tall trees could not be explained by this theory.

(b) Root Pressure Theory

It is observed in some plants that when they are beheaded (= cut off) near the ground level, water flows out through the cut surfaces. This water is sap of the xylem and the phenomen is called *exudation*. In grape vine, this is observed and is called *bleeding of grapevine*. The xylem sap is forced (= pumped) outside the plant due to the hydrostatic pressure which develops in the roots. The pressure which lifts the sap upward in the plant is actually due to the force exerted by the roots and hence, it is called *root pressure*. It should be noted that root pressure is an osmotic phenomenon. Although root pressure lifts the water in the plant to some height, the pressure (= force) is not so high that it could move water upward to very tall trees like eucalyptus, red wood *etc.* Further, the phenomenon is not observed in case of all plants and hence, is not supported.

(c) Physical Theories or, Physical Force Theories

These theories are based on the principle that physical forces are responsible for ascent of sap and the phenomenon could be demonstrated even if the cells are *dead.* The different theories under this concept are as follows.

Atmospheric pressure theory is one which has been put forward by Boehm (1809) and others. But this is not supported owing to the fact that 1 atmospheric pressure can lift water only upto 9–10 metres and thus, 50–60 atmospheric pressures would be required to lift water in a tall tree. Besides this, it may be stated that atmospheric pressure is only static and not continuous. To lift water, 10–12 atmospheric pressure may continue and further pressure would involve to overcome friction.

Imbibition theory is another which had been put forward by Sachs (1878, 1879). It is based on the principle that imbibitional role played by the cell walls of xylem brings about ascent of sap. Although the postulation got some support but had not been fully agreed as it has been observed that if the vessels in the cut out part of stem are sealed by paraffin or such other substance, wilting (= drying up) of shoots results and ascent of sap does not take place.

Some plant physiologists have brought to light *capillary concept* to explain ascent of sap. According to them, water moves inside the lumina of vessels and tracheids by capillary force. But this does not get support as water could be lifted only upto few centimetres if the vessels are large but will be much less if the vessels are of smaller diameter.

Cohesion theory is put forward by Dixon and Joly (1894). This is based on the concept that the water column in a plant body maintains a *continuity right from the leaves and upto the roots.* Thus, when there is a pull from the leaves, the pulling force extends down to the roots and the force that is exerted in the roots cause the roots to draw water from the soil. Thus, water is absorbed by the roots as a result of *pull from the leaves* and not due to the push exerted by the roots. The pull is produced by the leaves as a result of loss of water from them and it should be noted that the loss of water in the leaves takes place due to the physiological phenomenon which is known as transpiration. Thus, *pull due to transpiration in leaves is responsible for ascent of sap* from the roots.

The phenomenon is actually controlled by three factors. These are, (i) water has a rigid cohesive force, (ii) water column in the plant is continues in the plant body without any disruption and (iii) the pull, *i.e.*, tension that is produced takes place due to transpiration.

Cohesion refers to the fact that molecules of water are firmly attracted among them and in such a way that it becomes difficult to separate them. *Adhesion* denotes that water molecules are firmly adhered to the walls of the xylem vessels as droplets of water are adhered on the walls of a glass which contains it and when the water is taken off from the glass. For such strong cohesive and adhesive properties of water, the continuity of water column which exists in the xylem duct is difficult to be broken.

When transpiration proceeds, water in vapour form goes outside into the air from the mesophyll cells of leaves. Due to that, the *diffusion pressure deficit* of water in the cells goes high. Water is absorbed by the cell-wall from the cell sap, *i.e.*, vacuole. For the increase of diffusion pressure deficit that takes place, the cells take up water from the mesophyll cell. The mesophyll cells lying on the border of the vascular bundles leave their water and take up water from the xylem vessels of leaves. Increase takes place in the diffusion pressure deficit of the xylem vessels. The whole process exerts pull from the roots. Instead of going further details of the 'cohesion–adhesion–transpirational theory', it may be concluded that this theory of ascent of sap is *widely accepted now*, although there are some shortcomings in it and criticisms have crept in.

Downward Translocation of Water

It should be noted that water not only moves in upward direction in the plant body but is also moved downward. Dixon (1924) demonstrated this in potato plant. He observed that when the tip of a leaf in a potato plant is cut off in a solution of eosin (= red coloured dye), the solution moves downward through the xylem, petiole and is ultimately translocated to the potato tubers which remain underground. Red dye is seen in the tubers.

Lateral Movement of Water

It is observed that lateral movement of water also takes place from cell to cell in stems of plants. This takes place in a radial way. Experiments done in trees of woody species have demonstrated lateral movement of water in roots.

be, it is the costliest of all agro–inputs, less availability of water resource, high labour cost and perhaps due to fact that the necessity of irrigating all crops is not so much felt. It is estimated that not more than 280 million hectares of area in the world receive irrigation water at present and this accounts for 20–25 per cent of the total land under cultivation.

Largest irrigated area is now in China where more than 690 kilometre square is irrigated which comprises about 21 per cent area of irrigated land. India's position is next to China and USA is in the next position where irrigated area comprises 264 kilometre square. Irrigated area in Pakistan is also considerably high in which country, more than 200 kilometre square are irrigated.

Irrigated area in India is 140 million hectares which comprises 48.8 per cent of agricultural land. In terms of percentage, Punjab ranks highest where it is 98.1 per cent and other important States are Haryana, Uttar Pradesh and Andhra Pradesh where 87.6, 75.9 and 63.9 per cent areas are respectively irrigated.

Major sources of water for irrigation in India comprise water that flows on surface and also that which seeps into the ground. Canals are the primary source of irrigation in India. (Largest canal is Indira Gandhi canal which is about 650 kilometres in length. West coast canal in Kerela and Sethusamudram canal in Tamil Nadu and Kerala are also long canals). Apart from canals, tanks and wells are also important source of irrigation in the country.

Canals originate from the rivers. The north Indian rivers originate in the Himalayas and hence, being snow–fed, have less fluctuations in seasons. Seasonal fluctuation is high in the central and south Indian rivers which depend on monsoon, for which flow is maximum during rainy season. High attention has been paid to increase area under irrigation in the country by construction of canals, dams, barrages, deep and shallow tubewells and other projects. Irrigation projects in India are grouped as major, which cover more than 10,000 hectares of catchment command area, medium which cover 2000–10,000 hectares and minor irrigation projects which command less than 2,000 hectares of catchment command area. Water resource day in India has been marked on the eleventh day of March. (March 22 is regarded as the World Water Day).

Irrigation in Fruit Crops

Although India ranks second position in the world in regard to area under irrigation, much attention is paid in the country in irrigating specific crops and notably, the food grains. Rice is the dominant crop in the country and 30 per cent of irrigated area comes under this crop. Area under fruit crops in the country is estimated to be 6480,000 hectares but irrigated area under fruit crops could be said to be abysmally low. Even mango which is the most important fruit crop and occupies about 20,000 hectares area of land does not receive adequate irrigation. Some fruit crops which are given relatively much attention to irrigate in India could be cited as banana, grape, papaya, pineapple, strawberry, melons, apple, pear, sour lime, sweet orange *etc.*

Non–application of irrigation in fruit crops could in many cases be said to be a belief amongst the fruit growers in India that irrigation is not required in fruit culture. Such a concept comes in their mind from the general observation that many fruit trees grow well and produce satisfactorily without application of any artificial irrigation and hence, irrigation is not an essential feature for growing fruit crops. In fact, it is although noticeable that many fruit trees grow and produce without any artificial irrigation but this is due to the fact that they send forth branches of roots to very long distance in search of water and nutrients and get their requirement. This does not mean that they do not require water and such a thinking is only an absurdity. It should be remembered that requirement of water is also not the same for fruit plants belonging to all species. That is, requirement is less in some species and more in other species. There may even be difference in the requirement of water in varieties belonging to the same species of a fruit plant.

Chapter 5

Factors Affecting Irrigation Requirement of Crop Plants

A number of factors are known to play influence on the amount, time and frequency of irrigation water which is to be applied to a fruit crop. Some of the important major and minor factors have been discussed in the following.

1. Pant Factor

The amount of irrigation water which is to be applied largely depends on precipitation and the amount of water that is utilized by the same crop. Shanmugavelu (1987) in his book, "*Production Technology* of *Fruit Crops*" has grouped fruit crops into the following three categories in respect of their requirement of water.

(i) Those fruit crops which necessitate a permanent system of irrigation because without an assured supply of water, they will not be able to survive or will fail to produce economic crops, *e.g.*, banana, citruses *etc.*

(ii) Those that can be grown with *supplementary irrigation* at times of stress, *e.g.*, stone fruits.

(iii) Those that can do well under rainfed conditions, *e.g.*, mango, cashew *etc.*

The above classification appears to be befitting in tropical situation in south India and the eastern India. However, some sort of irrigation has its necessity for these fruit crops in most parts of northern India which is due to inadequate rain.

The species of the fruit crop is by far, the most important factor that determines requirement of water. Thus, fruit crops like banana, sapota, citruses, litchi, grape, apple, pear *etc.* may be mentioned which require higher amount of water as compared to fruit crops like custard apple, *ber phalsa,* date, emblica (*aonla*) *etc.* which have considerable drought resistance capacity and accordingly, they require relatively less amount of water. Mention may be made to dragon fruit, which is a fruit plant belonging to the cactus family and requires negligible water. Requirement of water may even vary in varieties, sub–species, hybrids *etc.* within a species.

Some species of fruit plants make excellent economy of water and hence, require less water from outside source. Pineapple is a classic example of this group. Leaves of this fruit plant are concave and thus, rain water when falls is conducted to the base. Leaves are compact and rosetty for which water is stored in their axils for considerable period. Stoma are few and they are only on the under surface of leaves. They are even covered by umbrella like projections which are known as *trichomes* with the help of which, they retard transpirational loss of water. Leaves are waxy for which evaporational loss of water from them is negligible. Roots are not much and they abound near the surface soil. Thus, the plant takes less water from the soil for these contrivances but it deserves interest as to why the plant does its best with high fruiting performance when grown in the high rainfall, tropical zone.

Root system of a fruit plant is beyond doubt, a pertinent factor that determines its requirement of water. Feeder (= absorbing) roots of some fruit plants abound mostly on the surface soil. These shallow–rooted fruit plants, *e.g.*, papaya, litchi, banana, sapota, citruses, grape, avocado, pineapple, strawberry, guava, peach *etc.* are benefitted much when irrigation water is applied more frequently and in lesser amount as the surface soil often gets dried. On the other hand, heavy irrigation is of help to the deep–rooted fruit trees like mango, jackfruit, jamun, fig *etc.*

Die–back, wilting, leaf fall, flower and fruit drop are some of the common symptoms that result in fruit trees when there is stress of water. A fruit tree mangosteen is observed to produce scanty root–hairs even in slight stress of water.

Age of the fruit tree is another determining factor for the rate of water uptake by it and thus, the necessity of irrigation. In young stages, fruit trees do not produce adequate roots and root–hairs are confined to the upper layer of the soil. Hence, at the early stage of growth, irrigation water should be applied in lesser amount but with greater frequency.

Application of irrigation water should also depend much on the vegetative and reproductive stages of growth of a fruit plant. For example, at some stages like dormancy of buds, flower–bud differention and initiation, pre–flowering *etc.*, requirement of water of the plant is less. On the other hand, greater necessity of water is felt at some stages, such as opening of vegetative buds, emergence and growth of vegetative flushes, fruit set and fruit development. Stress of water at these stages produce undesirable symptoms like withering of buds, stunted growth of shoots and inflorescences, wilting of leaves and their drying along margins, die–back of shoots, shedding of flowers and fruits. Some fruits like grape, litchi, citruses, sapota *etc.* undergo shrinkage and drying of pulp while in some fruits like

grown up condition. Accordingly, the requirement of irrigation is increased when close planting is done. Frequent tilling of the soil causes more aeration to the soil particles and for this practice also, uptake of water in increased and thus, necessity of irrigation may be felt. By exposing lower layer of soil, which is usually done with a soil–turning hoe, loss of soil moisture is increased at still higher rate and may necessitate higher irrigation.

Uncontrolled and luxuriant growth of weeds in the plantation take up huge amount of water from the soil and irrigation requirement becomes higher in such a plantation.

The practice of mulching on the orchard floor as described in the later part of the book is an efficient soil management practice if properly done. This effectively conserves soil moisture, prevents growth of weeds and as a result, application of water by irrigation is dispensed with or the amount and frequency of irrigation might be reduced to a considerable extent in the orchard.

Mazumdar, B.C. in his book, *Principles and Methods of Orchard Establishment* (2004) has stated that effective windbreak rows of trees greatly reduce the velocity of wind and direct the wind to blow in upward direction. As a result, evaporational loss of water from the soil surface as well as transpirational loss of moisture from the leaves are greatly reduced. Irrigating water to the orchard could therefore, be greatly reduced by the provision of suitable windbreak.

When intercrops are grown in the orchard, irrigation requirement evidently should be greater in order that competition does not take place between the fruit plants and the intercrops in respect of soil moisture.

Training and pruning practices that are adopted in fruit trees play much influence on the requirement of irrigation in the orchard. It may be stated that training fruit trees by open–centre (= vase) system involves greater requirement of soil moisture than the central leader system. Similarly heavy pruning particularly by heading–back system reduces the requirement of water.

Application of manures and particularly, chemical fertilizers has great influence in regard to frequency and amount of irrigation which is to be applied in the orchard. Unless there is heavy downpour, irrigation should be a must just after application of any fertilizer. Irrigation water brings the fertilizers (= salts) into solution form, subsequent movement of the solution in the soil, lowering of its osmotic concentration and thereby, uptake of the solution (= fertilizer, salt solution) by the root–hairs of the roots. Amount of irrigation water that is to be applied after fertilizer application would depend upon the type of fertilizers and their amount that are applied. In this connection, it may be mentioned that when the fertility status of the orchard soil is low, the roots of the fruit plants have a tendency to take up large amount of water which is done by them to derive from the soil the nutrients that are required by the fruit plants.

Apart from fertilizers, other chemicals that are applied on the orchard soil or foliar application of them which is done in solution form directly on the fruit plants exert great influence on the requirement of water. Few instances of chemical application are stated in the following:

Soil amendments are applied to ameliorate the acidic or alkaline soil by supplementing required amount of soil–amendments, *e.g.*, lime, calcium phosphate, basic slag, gypsum, pyrite, sulphur, molasses *etc.* Just after application of these or other soil ameliorating agents, heavy irrigation no doubt, becomes necessary.

Anti–transpirants are much used nowadays in many fruit crops. These reduce the transpirational loss of water from the plants and thereby, reduce the requirement of irrigation. These agents may be physical, chemical or hormonal in nature. Physical substances are colourless plastic materials, silicone oils and low viscosity waxes and these are sprayed in solution form or dusted over plants and due to that, a thin film is left on the leaves and plant surfaces. This film prevents moisture to escape and transpiration is reduced.

Among chemical agents, a number of organic compounds which are rich in carbohydrate are applied on plants. Sugar solution is also sprayed as an anti-transparent. These chemicals liberate carbon dioxide gas and thereby, air layer surrounding leaves and plant surfaces are enriched with carbon dioxide to about 0.05 per cent. As a result, transpirational loss of water is reduced and makes reduction in requirement of water.

The chemical, Phenyl Mercuric Acetate (PMA) which has also fungicidal property is a promising type of anti–transparent chemical and is much used now for high effectiveness, less toxicity and low cost. Spraying aqueous solution of this chemical to about 10^{-4} (M) concentration induces partial closure of the stoma for two weeks or more in many crop plants. Another chemical, *viz.*, Triadimefon, which has also fungicidal property is used in many plants.

Plant growth regulating chemicals are sometimes applied as foliar spray solutions to retard transpirational loss of water from leaves. Among these compounds, auxins such as NAA, NAd, beta–NOXA *etc.*, chlorine-substituted phenoxy compounds, *e.g.*, 2, 4–D, 2, 4, 5–T, 2, 6–D *etc.*, are used besides the notable growth regulating chemical, *viz.*, abscisic acid which is a highly effective anti–transpirant. However, the optimum concentrations of these compounds for high effectiveness are found to overlap with toxic concentrations of them.

In viticulture, spraying the grape vines with polyethylene type, or a vinyl acrylate type of materials is done in some countries for the purpose of delaying harvest of berries by 2–3 weeks. These chemicals also cause reduction of water stress in the berries for which irrigation requirement is reduced. These have also been found to cause about 30 per cent reduction in water uptake by the plants. On the other hand, some chemicals are there which make increase of transpiration from leaves by their applications and thereby, necessitate irrigation. Among other chemicals belonging to this group, copper fungicides including Bordeaux mixture may be cited as examples.

Chapter 6

Determination of Soil Moisture Content

Knowledge of the existing moisture content of the orchard soil is necessary in order that the irrigation water can be applied in a judicious manner. Soil moisture content could be assessed by a number of methods, each of which has merits and demerits. The common methods are stated below in brief:

(a) From Personal Experience

In a country like India, crop culture and fruit farming run in families. The farming population learn many sophisticated technologies of cropping procedure from elderly persons in the family and without undergoing study in an agricultural or a horticultural college. There are many so expert crop growers all over the country who just seeing the soil or trampling with feet or hand feeling of soil sample can presume the moisture content of soil. The methods which are commonly adopted by them may be stated as follows:

(i) Crumbling of Soil Samples

The fruit grower takes samples of soil from the surface and from greater depth and also from different parts of the orchard. The samples collected are separately crumbled by the fruit grower for sometime and sometimes these are crumbled taking inside cloth pieces. On doing so for sometime, he is able to assess moisture content of the orchard soil although not mathematically and he decides whether the orchard needs irrigation or not.

(ii) Examining Leaf Samples

In this method, the orchardist tears some leaves along with petioles from the fruit trees of the orchard. Then he examines sap of the leaves and the sap at the base of the petioles. By such careful examination, he can assess the moisture content of the fruit trees, *i.e.*, whether irrigation should be given or not in his orchard.

It may be stated in this connection that in Hawaii in the USA, the pineapple growers (pineapple is abundantly grown in Hawaii) assess soil moisture content and necessity of irrigation in the plantation by cutting across a leaf from the fourth whorl back from the centre of the plant. If the palisade layer of cells (= elongated, vertical cells lying just below upper surface of the leaf) is not about half the thickness of tissues lying under this layer, it is considered that irrigation is necessary.

(iii) Examining Sap of Leaf–sheaths

In banana, the pseudostem (= false stem) which is formed by clasping together of the leaf–sheaths is punctured and the sap is observed to determine the necessity of irrigation.

(iv) Examining Curling of Leaves

In some fruit trees and especially in citruses, if the leaves are found to be curled, it is presumed that the soil moisture content has reduced and irrigation is considered to be needed in the orchard.

(v) Drying of Leaf Margins

In papaya, grape and other fruit trees, leaves are very thin and the margins have lobings. Hence, dry appearance and brittleness of the leaf margins are reliable signs that the fruit plants are in water stress and necessitate irrigation.

(vi) Drooping of Leaves

Deficiency of soil moisture and necessity of irrigation in some fruit trees are known if the leaves or the twigs containing leaves bend downwards, *i.e.*, they droop.

(b) By Determining Moisture Content of Leaves

Especially for thinner–leaved plants, water requirement may be assessed fairly well by examining the moisture content of their leaves. Simmonds, Samuels and other plant physiologists suggested that like sugarcane, in banana also, among "accessory methods of indicating water requirements, leaf moisture would appear to offer the best chance of providing a rapid and reliable indication".

The methods which have been stated above are practical and do not involve cost. But they are not always very reliable and sometimes even misleading. Where moisture content of soil is needed to be precisely known and for carrying out irrigation experiments, more sophisticated techniques and instruments are nowadays used. There are many methods and many instruments are devised regularly for the purpose. Some of the common methods are stated in the following.

(c) Gravimetric Method

Soil moisture which exists in free state is determined in this method in a direct way. To do so, soil samples are collected from different locations as well as different depths of the concerned orchard. This can be easily done with a handy, pipe–like device known as *soil–auger*. The soil samples which have been dug out are separately weighed to determine the initial weight (W_1). Then the samples are dried in a hot–air oven for 60–70 hours and maintaining a temperature of 100°–105°.

After that, *i.e.*, when the samples are dried, they are taken out of the oven and weighed without making delay, *i.e.*, while they are still warm. This weight gives the final weight (W_2). The difference between the initial weight (W_1) and the final weight (W_2) indicates the loss of water in the soil sample, *i.e.*, evaporational loss of water during drying. The loss of water is expressed in terms of percentage. That is, it is expressed as amount of water (by weight) which is lost if the sample weighs 100 gram. However, instead of weight basis, loss of moisture is also determined on volume of water basis also. Where there is no facility of drying in oven, soil samples can be dried in organic solvents like ethyl alcohol or methyl alcohol, rectified spirit, benzene *etc.* but calibration becomes necessary with gravimetric methods if drying is done by organic solvents.

(d) Use of Instruments

Under field–scale, *i.e.*, directly in the orchard without taking soil samples, soil moisture content may be known on the spot by the use of some instruments. A number of instruments are used for the purpose and among those, *tensiometers* and *resistance blocks* are commonly used. These instruments are however, needed to be calibrated to the particular condition before they are used and they give precise results.

Tensiometer works on the principle of measuring the force with which the soil pulls the water. There are various types of tensiometer and the simple type consists of a spongy cup with a neck, which is fitted to a rubber tube and the open end of the tube is attached to a mercury manometer or a vacuum gauge. The tube is filled with water and the cup of the tensiometer is placed in the depth of the soil where moisture content is to be determined. Movement of water takes place between the water in the cup and the soil water such that, an equilibrium is reached. The high or low tension which is produced is recorded by the manometric reading and this indicates the soil moisture content. Tensiometer consisting of many cups is also available, which can simultaneously determine soil moisture content at different layers of soil.

Where soil moisture content is much lower, tensiometer is not very effective and in such situation, resistance block is more efficient and is also simpler to use. It consists of a block which is made up of a porous medium such as porcelain, gypsum, nylon *etc.* with which electrodes are connected. The block is placed in the desired depth of the soil and the electrodes are fitted to the potentiometer. The moisture content of soil reflects changes in electrical resistance and is recorded in the meter. For interpretation, resistance block is calibrated against the gravimetric method of determining soil moisture content.

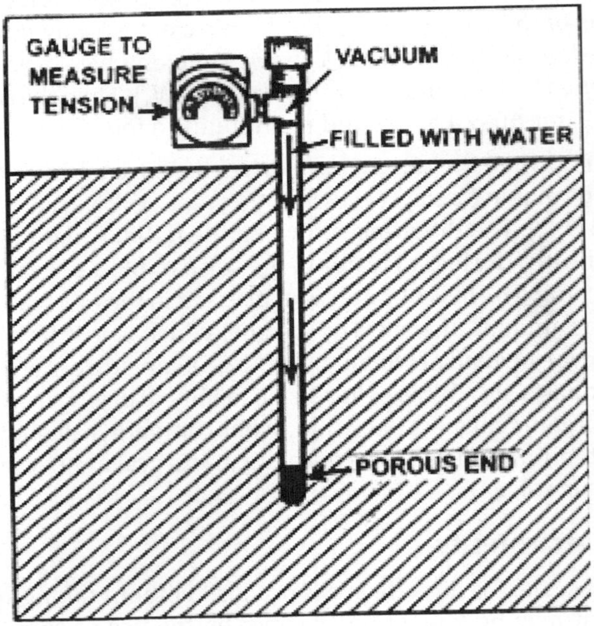

Figure 1: Tensiometer.

Electronic soil moisture detectors are much on use now. These are portable, battery-operated devices and are designed for rapid and simple testing directly in the fields or orchards. They read directly from 0 to 100 per cent available soil moisture. The instrument of this type consists of a gypsum block which is permanently embedded in soil, electronic circuit, a moving coil instrument calibrated directly in moisture percentage from 0 to 100 per cent, standardizing knob and input terminals. The soil block is put in contact with soil for 24 hours or more before measurements are made. Various types of such instruments are available.

Soil moisture sensors of many other types are also available, Neutron moderation method is one of the precise methods of determining soil moisture content. This is of much help particularly when the soil contains high salt concentration. The method involves emitting of radio–active neutrons at a high speed from a radium source which is installed inside the soil. The emitted electrons collide with hydrogen atoms present in soil moisture. By such repeated collision with hydrogen atoms, speed of the neutrons is slowed down as they loose energy and their direction of movement is also changed. Then the slowed down neutrons strike on a suitably placed silver or rhodium plate and make it radio–active. If the soil contains more water, *i.e.*, hydrogen atoms are greater in number, number of slowed down neutrons will also become greater and accordingly, the metal plate will acquire greater radio–activity. Hence, radio–activity of the plate is measured to determine moisture content of soil.

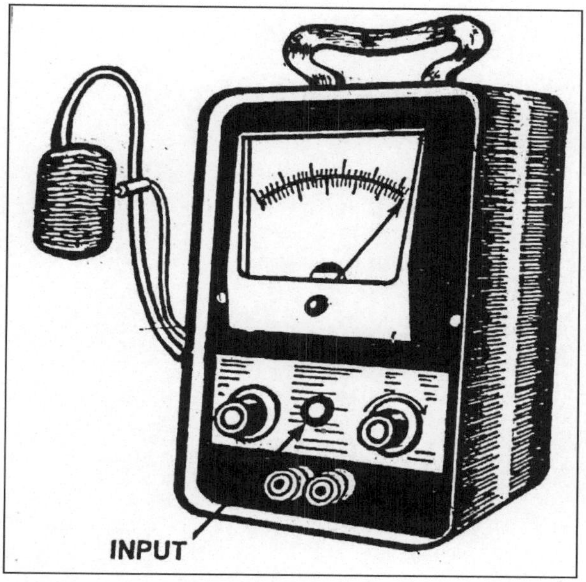

Figure 2: Electronic Moisture Meter.

Frequency or time domain reflectometer, galvanic cell *etc.* are also used. Thirsty light sensor which emit blinking light in dry soil and determines moisture level very frequently is another device.

Chapter 7

Principles of Irrigating Fruit Crops

Leaving aside the fruit crops which could be grown under rain–fed condition, irrigation is no doubt, a basic necessity in farming of fruit crops in order to secure optimum yield and better quality of the produce. But it should be remembered that water management in a fruit crop should be done judiciously keeping in mind a number of considerations, because improper and faulty irrigation make more harm than any good. The following principles among others are of special significance in irrigating fruit plantations.

Economic Use of Irrigation Water

Irrigation is the costliest input in agricultural and horticultural farming. Hence, the operation should be done in a justifiable manner and ensuring economic measures. Even water itself cannot be said to be very cheap today and use of the idiom, "as cheap as water" has limitations. The United Nations Organization predicted that 7 billion people in the world might experience water shortage by 2050. A report states that if the current growth rate of population in India which is 2.11 per cent per year now continues, the population in this country which comprises 17 per cent of the total world population will exceed China within a decade and there might be a severe crisis of water. Hence, wastage of water is not justifiable by any means and "cold water should be thrown" on them who waste water.

It is noticed in many orchards that irrigation water sometimes goes into wastage through leakages, such as rat holes, crevices, tunnels *etc.* To prevent such loss, release of irrigation water from the source alone is not enough. In such case, the orchardist should also occasionally check up the course of irrigation water in the orchards and make suitable arrangements to seal the leakages wherever these are detected. Loss of water due to leakages in the channels may however, be saved by delivering irrigation water through pipelines, hose–pipes, polythene tubes *etc.* Pipelines should be of wide diameter and not very narrow. Lining the inner sides of the pipelines with long polythene sheets prevent loss of water due to leakage. It should be ensured that hose–pipe, polythene tubes or rubber tubes, if used for conveying irrigation water should not have any hole in any part.

Careful Irrigation in Shallow Rooted Fruit Plants

In case of shallow rooted fruit plants *e.g.*, papaya, litchi, guava, pineapple, strawberry, avocado, sapota, peach *etc.*, majority of the *feeder roots*, *i.e.*, those roots which take up water and nutrients from the soil abound near the upper layer of soil, *i.e.*, within about 10 cm from the surface. When there is high wind, high atmospheric temperature or after heavy rain, the top soil is displaced and thus, wasted. As a result, the feeder roots are exposed, come to the ground and water and nutrient uptake by the trees are seriously hampered. Even under normal condition, the top soil often loses moisture and thus, water stress is felt by the plants which remains unnoticed to the orchardist.

Water management of the shallow rooted fruit plants is best done by irrigating them at shorter intervals and applying less amount of water during each time of irrigation.

Irrigation at Specific Stages of Growth

Stage of growth and development of a fruit plant is a very important consideration to irrigate it to derive the real benefit. In fact, irrigation has great necessity in certain stages of growth while it may cause harm if applied in some stages.

Requirement of irrigation is felt in the following stages in particular:

(i) Emergence and Growth of New Shoots

In evergreen trees, large number of leafy growths appear in several seasons in the year which are known as *growth–flushes*. In deciduous trees, new growths appear only once in a year in spring after leaf fall in winter. To ensure proper growth of the newly emerged twigs, adequate moisture should be present in the soil, failing which the twigs may be wilted. Hence, irrigation should be a must if there is deficiency of soil moisture at the time when the twigs have sharted to grow.

(ii) Fruit Set and Development

Moisture stress during setting of fruits is a major factor for high drop of fruits. But irrigating a fruit plant when it has set fruits has also an adverse effect in that, fluctuation of moisture content in the pedicels may cause high drop of fruits which have just set. Hence, to ensure proper moisture level in the reproductive parts, it is advisable that irrigation should be done well in advance before inflorescences have appeared.

Relatively high amount of moisture is necessary to be present in a fruit tree when the newly set fruits have started developing. Moisture content in the fruits should be in optimum level till they are fully mature. Shrivelling or misshapen fruits due to deficiency of moisture is a very common occurrence in some fruits, *e.g.*, grape, litchi, longan, custard apple, bullock's heart, jackfruit *etc.* Moisture stress during fruit development brings about cracking of many fruits and some examples are litchi, longan, grape, citruses, bael, pomegranate, guava, kiwi, *ber*, cashew nut *etc.*

Heavy dropping of premature fruits is a common abnormality of moisture stress in fruits during their development. Some examples are mango, citruses, papaya, litchi, *aonla*, sapota, bael, jamun, coconut *etc.* Although natural drop of fruits at some stages of fruit development is a common occurrence in some fruits, *e.g.*, "early drop", "June drop" and "pre–harvest drop", in apple, "pin–head drop", "post–setting (April) drop," and "May drop" in mango *etc.* but the intensity of fruit fall becomes greater in moisture stress of soil. Hence, liberal irrigation in the orchard at suitable intervals during development of fruits has its necessity to prevent the loss of fruits during their premature condition.

In fruits belonging to a large number of species, high moisture content of the fruits during their ripening condition is not considered desirable. In grape vine in particular, irrigation is completely stopped about a month before harvest in some parts of Andhra Pradesh.

(iii) Early Stage of Growth of Fruit Trees

Roots of those fruit trees which live long and attain maturity for flowering (= gestation period) after attaining few years of growth do not spread much and their feeder roots are present near the soil surface. Hence, like shallow rooted fruit plants, their roots (= root–hairs) often suffer from water stress. To avoid moisture deficit, these fruit plants need irrigation and this should be applied to them at shorter intervals and in relatively less amount. Spraying water to the young fruit plants which may not be in fine droplets is also a good practice along with light irrigations. Throwing water on the above ground parts with a mug is also a good practice.

(iv) Irrigation after Application of Fertilizers

Just after any type of fertilizer which is applied to the orchard soil, it should be thoroughly mixed with soil with suitable hoeing implements and then copiously irrigated. It is mentioned in this connection that superphosphate is not easily broken down to pulverized form and remains as small clots. Hence, it is advisable that it should at first be ground as far as possible in fine form and then applied to the orchard soil. (If green manuring is done in the orchard, superphosphate is best applied in green manuring plants which has been discussed in the second part of the book).

Many horticultural scientists are of the opinion that, if heavy amount of fertilizers are applied, two irrigations should be applied instead of one. That is, after the first irrigation, second irrigation should be applied after waiting for 5–7 days. The idea behind this practice lies in the fact that, heavy irrigation just after application of chemical fertilizers may cause washing of some fertilizers to distant place away from the root zone of fruit trees.

Hence, the first irrigation should be given rather lightly and the second irrigation after 5–7 days and heavily.

(v) Irrigation after Application of Soil Amendments

After application of chemical soil–amendments, *e.g.*, gypsum, lime, sulphur, basic slag, calcium phosphate *etc.*, irrigation is essential after thoroughly mixing them in the orchard soil. It should be noted that gypsum is not much soluble in water. Hence, it should be applied in advance and preferably, 2–3 irrigations should be applied after gypsum application to ameliorate alkaline soil to bring to desired pH level.

(vi) Avoiding Irrigation after Manuring

Just after application of organic manure in the orchard soil, it is not advisable to apply irrigation in most cases. If the soil is very hard such that mixing the manure with soil becomes difficult with hoeing implements, light irrigation may be done and this should be done before application of manure. After waiting for few days when the soil comes to workable condition with implements, manure should be applied and thoroughly mixed with the soil.

(vii) Water Spray after Dust Storm

In some parts of India, dust storm appears to be a serious problem. Dust particles cover leaves and other parts of the fruit trees. In north India, such a storm is locally called *andhi* and in West Bengal, *kalbaishakhi*. In many other parts of India and in the coastal areas, dust or sand storms occurs in some seasons.

Due to dust storm, dust particles cover the leaves and accordingly, photosynthesis, transpiration and evaporation from the leaves are greatly hampered. As a result, uptake of water and nutrients from the soil are *severely* impeded. The dust particles adhere to leaf surfaces for several days.

To get rid of this hazard, irrigation should be given to enhance photosynthesis and dissipate moisture from leaves by evapo–transpiration. However, the best method is to bathe the trees by *complete coverage spray with water* and due to that, dust particles are washed out and stoma will be opened.

(viii) Warming up of Orchard Soil

In temperate zone in particular, the soil temperature sometimes goes very low in cold weather. Dropping down of soil temperature to very low level causes severe damage to the roots of the orchard and even ice formation in the protoplasm of root cells.

In such situation, *i.e.*, when soil temperature goes to very low level, one or two flood irrigation in the orchard is found to be of much help in raising the soil temperature to some extent. This happens due to the fact that micro–organisms which inhabit in the soil become activated in getting water. As a result, their rate of respiration, *i.e.*, anaerobic respiration increases. Thereby, they liberate energy, *i.e.*, heat–energy and makes somewhat increase of soil temperature.

Chapter 8

Irrigation System: Surface Irrigation

In accordance with *placement of irrigation water*, there are three systems that are followed in irrigating orchards. These are:

(i) Surface
(ii) Sub–surface and
(iii) Overhead irrigation systems.

Each of the system includes a number of methods, the suitability of which depends on certain conditions.

Under Indian condition, the surface irrigation system is beyond doubt, more common. In this system, water is delivered directly over the soil surface of the orchard. The common methods under this system of irrigation have been stated in the following.

1. Flooding

In this method, irrigation water is freely discharged on the orchard land in such a way that the entire soil surface is filled with it as if it has been flooded.

The first step to irrigate by this method is to *level the land as evenly as possible.* Besides, the bundings are also needed to be made at all boundaries of the land to hold the water in it. To irrigate the land, a water channel is made to open at any place of it and the irrigation water which is conveyed through this channel is collected on the land. The water thus, collected at one place spreads freely on all sides and ultimately, the entire land gets saturated, *i.e.*, flooded. When it has been done so, the entry of water is discontinued and the water that has been collected on the land percolates into the soil. If any excess water remains present after percolation, it is drained out through drainage channel made for the purpose or is allowed to evaporate naturally and slowly.

Irrigation by this method is possible if the size of the land is small as it is usually difficult to saturate the entire surface area where the area is considerably large in size. In the latter case however, water is to be let out at several places instead of restricting to collect at one place.

(a) Advantages

There are some advantages of flood irrigation method, which have been stated in the following:

(i) The entire soil gets fully saturated with water without leaving any dry pocket in it.

(ii) The method has its suitability where the crops require only a few irrigations and at long intervals, *i.e.*, where occasional irrigations are only given.

(iii) The method is sometimes preferred where interculture with heavy implements becomes necessary.

(iv) The method no doubt, finds favour where the land is tolerably flat, the soil is fairly compact in structure and above all, irrigation water is available plentifully and with negligible cost.

(v) As involvement of labourer is less, the operational cost is also less in this method.

(b) Disadvantages

The disadvantages are perhaps more in this method, particularly when the area is large and the notable ones may be stated as follows:

(i) The orchard trees badly suffer due to over–irrigation if the irrigation water is not drained off quickly and timely.

(ii) Especially where the clay content of the soil is higher, water stagnates over the land for long time and this may result even in death of the trees.

(iii) The method cannot be used unless the land is even.

(iv) Heavy watering by flooding very often results in luxuriant growth of weeds on the land.

(v) A huge wastage of water takes place in this method due to surface run–off, seepage, evaporation *etc.* and as a result, not more than 15 or 20 per cent of the applied water is only of real benefit to the orchard trees.

2. Check or Bed Method

This method is a modified and somewhat improved form of the flooding method, where saturations of the land with water is done part by part.

Like flooding method, in this method also, the first step should be to level the entire land as thoroughly as possible and bundings should be done at all the boundaries of it. Then as shown in Figure 3, square or rectangular compartments, *i.e.*, beds are made in the land by bundings so that one or more fruit trees are enclosed in a bed. The size of the beds however, varies and is generally made larger in clayee soil. The channels run in such a way that one channel connects two adjacent rows of beds on both sides of it. Irrigation water is released in a bed until it is fully saturated. After that, the flow of water is disallowed in it.

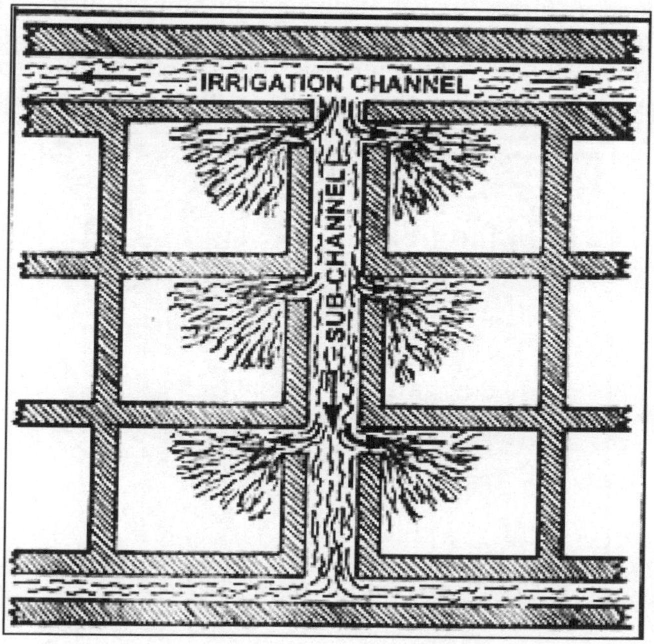

Figure 3: Bed Method of Irrigation.

(a) Advantages

The method has the following advantages over that of the flooding method:

(i) The distribution of water takes place more uniformly.

(ii) The chance of over–irrigation is less as the entry of irrigation water in the beds is regulated.

(iii) The soil is made to saturate more evenly.

(iv) The beds remain permanent and hence, recurring cost becomes relatively less.

(b) Disadvantages

Despite some advantages as stated above, the method does not gain much support in most places for the same difficulties as are faced under the flooding method that have been mentioned above under 1 (b).

3. urrow Method

This method consists of opening *long furrows* over the entire orchard area with the help of plough, harrow or suitable ridger. On opening the furrows, the irrigation water is made to flow *slowly* through the furrows (Figure 4). As a result, the underground soil as well as the soil on both sides of each furrow are soaked, respectively by downward and lateral percolation.

The efficiency of furrow irrigation depends on how the furrows have actually been opened on the land. That is, the number of furrows laid between two adjacent rows of fruit trees, direction of the furrows, their length, width, depth and other conditions are important factors that determine effectiveness of the method.

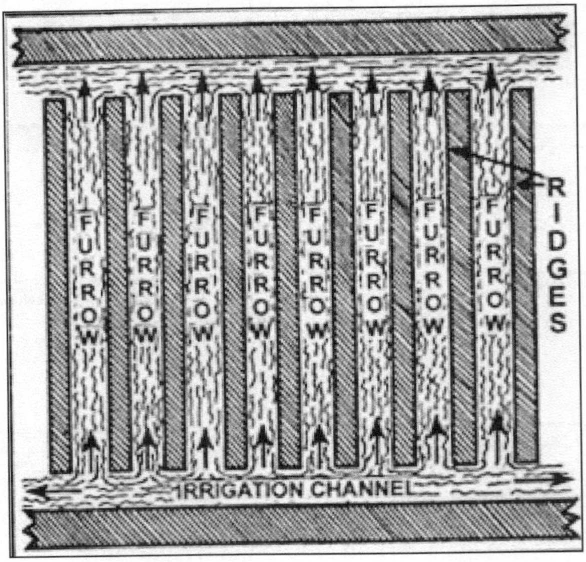

Figure 4: Furrow Method of Irrigation.

When the trees are young, their roots do not extend much and hence, only one furrow on each side of a fruit tree appears to be sufficient to irrigate them. At this stage of growth of the trees, maintaining many furrows between the rows is practically of no benefit. But as the trees grow, their roots cover more and more area and accordingly, the number of furrows between two rows also need to be increased. In this way, when the trees are in mature condition, the entire inter–row spaces are covered with their roots and accordingly, furrows should also be closer, *i.e.*, at 50–90 cm apart, with one furrow near to or under the end of the branches.

The furrows generally run in one direction. But opening furrows in opposite directions has its advantages as it facilities better distribution of water. Again, very long furrows are not useful because, loss of water at the heads of the furrows is greater in such long furrows and the trees lying near the mouths of them usually do not get adequate water. It may therefore, be stated that in general, more than 150–metre long furrows in lighter soils and more than 300–metre long furrows in heavier soils are not considered satisfactory.

As far as possible, the furrows should be wider and shallower because, such type of furrows make more effective distribution of water than the furrows which are opened deeply and narrowly.

It is also mentionable that water should be allowed to move in the furrows *slowly* and may even be left to stand in the furrows for some time. This facilitates better percolation of water and control of soil erosion. However, slow movement of water in the furrows is not always practicable.

(a) Advantages

The furrow method of irrigation has the following advantages over the flooding:

(i) Evaporational loss of water is less.

(ii) The water is distributed more uniformly.

(iii) The irrigation water is able to reach greater depth of the soil.

(iv) The method is convenient where distance between plants in a row is less as in hedge–row system of planting.

(v) For such fruit crops which ate needed to be earthed up occasionally as in banana, the method is preferable. This is because, earthing up is done automatically while opening the furrows.

(vi) Where the soil is of higher clay content, the method is considered to be particularly suitable. In such *retentive soil*, keeping part of the soil wet and the part adjacent to it as dry in an alternate manner is more efficient than drenching the entire soil surface as in flooding or bed which causes water stagnation due to poor percolation capacity of such soil and this is brought about by this method of irrigation.

(b) Disadvantages

Despite the above advantages, the following disadvantages are however, met with in this method:

(i) The method is not suitable in porous soil for high vertical percolation.

(ii) Many *feeder roots* lying close to the soil surface may be damaged or cut off while opening the furrows.

(iii) Hard layer may be formed under the furrows where the soil contains much clay. To break the hard layer, occasional sub–soiling may become necessary.

(iv) The furrows are very often choked by soil deposition and the soil needs to be removed for free movement of water which necessitates much labour cost.

4. Border Method

This method is also called as the "sara" method. To irrigate by this method, the land is divided into long strips by relatively broad ridges (Figure 5). The length of a strip may be upto 200 metres and the width may be upto 90 metres. The strips are levelled as far as possible. Each strip is then irrigated with a water channel unless it is thoroughly saturated.

The method is an improvement of the flooding method and less labour is in fact, involved to irrigate by this method.

Figure 5: Border Method of Irrigation

5. Basin Method

In this method, circular or square shaped enclosure known as *basin* is made by soil bund, separately around each tree and the irrigation water is let out into each such basin by means of inter–connecting channels.

To adopt the method, as shown in Figure 6, the basin is at first made around each individual tree with the help of soil bund. The diameter of a basin is an important factor and this is adjusted in such a way that its border, *i.e.*, the soil bund remains just below or preferably, slightly away from the skirt of the tree, in other words, the peripheral spread of its branches. Thus, at the early stage of growth of the trees when the peripheral spread, *i.e.*, the canopy growth is less, the size of the basins should also be less. Every year, the diameters need to be increased in consistence with sidewise growth of the trees. In this way, the increase in diameter of the basins should be continued till the trees grow to optimum maturity and this varies from 6 to 10 years, depending on species and variety of the trees as well as the soil, its fertility, cultural practices and the climatic condition. After attainment of this maturity, large increase in peripheral spread of the trees does not take place and accordingly, the basin size also need not be increased any more, which mean that the bunds at the boundary of the basins remain permanent.

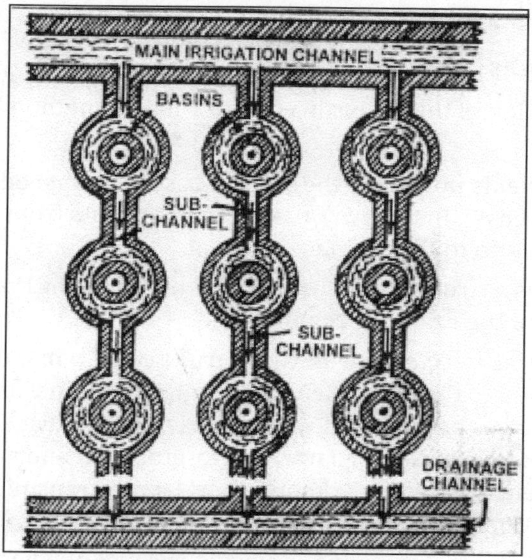

Figure 6: Basin Method of Irrigation

When the basins have been made around the trees, those which are in each row are inter–connected with sub–channels. To do so, the sub–channel which opens into the first basin is connected at its other end with the main channel. Similarly, the first basin is connected to the second basin with a sub-channels, the second is connected to the third basin with a sub–channel and in this way, all basins in the row are made inter–linked.

The drainage channel is then made at the other end, *i.e.*, near to the last basin in the same row. On making so, the last basin is connected to the drainage channel with a sub–channel. By this layout of basins and sub–channels, a continuity is thus, maintained from the first to the last sub–channel and finally to the drainage channel.

With the idea that a basin gets thoroughly irrigated, its sub–channel which connects the next is sometimes closed with soil and when it is almost flooded, the soil is removed to allow water to flow to the next basin. The same practice of blocking and opening are made respectively for each basin to ensure through drenching in it.

(a) Advantages

(i) The basin method of irrigation is helpful especially in loose–textured soil where flooding or furrow method has limitation. In such soils, making basin around a tree to irrigate it enables deeper percolation of water.

(ii) Water stagnation in a basin also does not take place as the excess water flows to the next basin and is finally drained off.

(iii) There is also much economy of water in this method as evaporational loss of water is less due to the fact that the water stands under the shade of the trees.

(b) Disadvantages

Inter–connection of the basins in each row in this method have the following drawbacks:

(i) The nutrients present in the soil in a basin are washed away and move to the next basin. In this way, part of the nutrients from all basins in a row ultimately go to the drainage channel.

(ii) The soil is also eroded from the basins and crossing the basins, ultimately it is left to the drainage channel.

(iii) The disease infected leaves, twigs, fruits *etc.* if remain fallen under a tree are carried to the other basins through irrigation water. Due to that, the healthy trees are infected with diseases by the diseased materials. The *anthracnose* disease in guava, papaya, grape *etc.* and many other diseases in different fruit trees are known to be largely transmitted by this method of irrigation.

(iv) The soil–borne pathogens, if exist in the soil of a basin are also carried to the other basins and thereby, the soil of those basins are also contaminated. The wilt disease in guava and some other fruit trees due to *Fusarium* and related fungi, virus diseases in some fruit trees *etc.* may also spread over large area when irrigated by this method.

6. Ring Method

The method is similar to the basin method except that the basins made around the trees are not linked together by inter–connecting channels.

Like the basin method stated above under the heading, (5), the basin is made around each tree. As shown in Figure 7, the main irrigation channels are then laid out between *alternate rows of trees.* The basins lying on both sides of each such main channel are then connected with it by sub–channels.

The drainage channels are also made between alternate rows of trees, *i.e.*, between those rows where the main irrigation channels have not been maintained. Each basin is directly connected to the drainage channels by means of sub–channels. Normally, the drainage channels become narrower than the irrigation channels.

However, the drainage channels are also omitted in this method by many fruit growers.

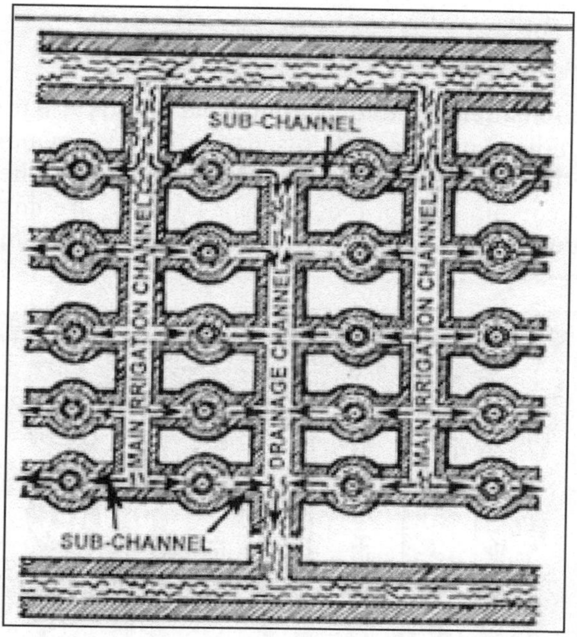

Figure 7: Ring Method of Irrigation

The basins made around the trees are thus, not inter–connected as in the ring method. When water is made to flow though the main channels for irrigation, it is delivered to the basins through the sub–channels which link these two. The excess water after irrigating a basin escapes to the drainage channel through the respective sub–channels. As stated in ring method, the water collected in a basin may be held for some time to enable better percolation. For this purpose, the sub–channel which links a basin with the drainage channel may be blocked with soil and after some time, the blockade is removed to enable water move freely to the drainage channels.

(a) Advantages

The disadvantages as have been mentioned above in the basin method under the heading, 5(b) are in fact, eliminated by adopting ring method of irrigation in an orchard.

(b) Disadvantage

If drainage channels are not maintained, the excess water that is left in a basin is not drained out. In loose–textured soil, although the water percolates into the soil gradually in such case, it is not so in soil of high clay and having poor percolation capacity. Thereby, the water stagnates in the basins and is evaporated rather slowly as it remains under the shade of the trees. Accordingly, in such type of soil where percolation rate is slow, the ring method is not suitable, unless the amount of water delivered to the basin is carefully regulated to avoid prolonged stagnation in the basin.

7. Combined Methods

Making a combination of the conventional methods is also preferred by many fruit growers in India to irrigate orchards, particularly when more than one fruit crop is grown in the same orchard land and more commonly when intercrops are grown. Although combination of the different methods are noticed in different parts, combination of the basin or the ring method with the furrow method is usually of occurrence.

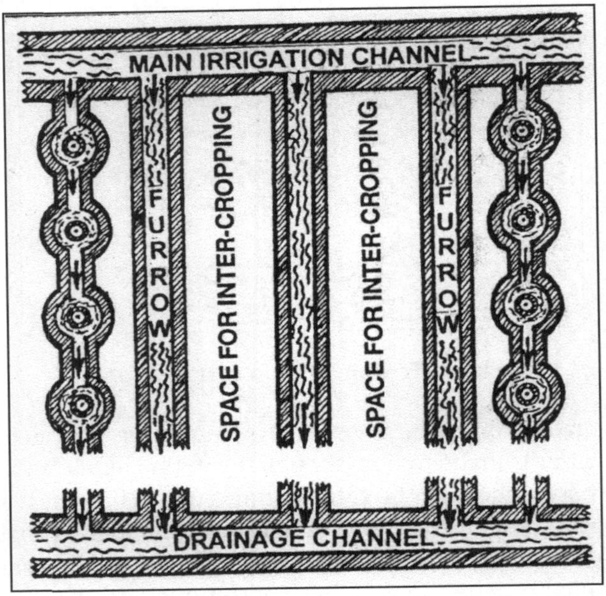

Figure 8: Combined Basin and Furrow Method of Irrigation.

When the basin method is combined with the furrow method (Figure 8), the furrows are opened in the interrow spaces. Much wide interrow spacing is in fact, left for the purpose of growing other fruit crops like banana, pineapple *etc.*, or inter–crops and widely spaced furrows are opened.

In combination of the ring and the furrow methods (Figure 9), two adjacent rows of fruit trees are maintained at normal spacing and the two adjacent rows are widely spaced for growing banana, pineapple or other intercrops. Widely spaced furrows are opened in the interrow spacing for irrigating those crops.

(a) Advantages

The methods are considered to be of advantage if more than one fruit crop or inter–crops are grown in the orchard. In such cases, the different crops appear to be irrigated by the methods which are most suitable to them.

(b) Disadvantages

Wastage of land is more in this method besides the fact that *high cost* which is involved. Over–saturation of the land may also result unless the discharge of water is regulated.

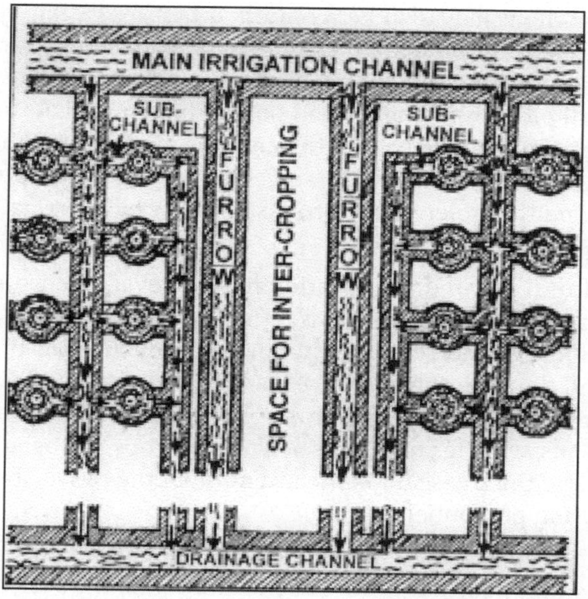

Figure 9: Combined Ring and Furrow Method of Irrigation

8. Drip (= Surface drip) Method

Drip irrigation, which is also known as *trickle–* or *micro–irrigation* is an innovation in the irrigational technology of the modern age which has made a breakthrough in irrigating agricultural, horticultural or plantation crops and forest trees in many parts of the world. The method has been widely adopted in many countries of the world notably, Israel, south and central America, Arizona, Taxas and other States of the USA, western Europe, northern Africa, Egypt, Australia, New Zealand, parts of India *etc.* and is irrigating an area of more than 4 lakh hectares. The area is however, fast increasing for the great benefit that is derived by this method. In India, the method has become popular in Maharashtra and Haryana.

As the term implies, drip irrigation involves delivery of irrigation water in very small amount, *i.e.*, by drippings (= in drops) and at a very low rate of flow on the soil mass lying close to the water–absorbing, in other words, feeder roots of the plant which is irrigated.

The method is therefore, based on the principle that only the required amount of water is supplied to a plant and in such a manner that every drop of water delivered reaches directly and readily to the soil closely adhering those roots which are actually concerned with uptake of water.

Although the present system of drip irrigation has been in practice since the sixties, the concept of it is believed to be an ancient one and there is myth that about 2600 years ago, Nebuchadnezzer used drip method for irrigating the "Hanging Gardens of Babylon". It may also be mentioned that irrigation by dripping is an age-long practice in the Hindu families of Bengal to water the sacred basil *(tulsi)* plants in the house during the summer months. This is done by hanging above the plants a small pitcher having a small hole at the bottom. A thread or a bunch of grass (*Cynoden dactylon*) is passed through the lower small hole of the pitcher which is filled with water occasionally and water is let out in drops at low rate. As these plants can neither tolerate moisture stress nor over–irrigation, the method is of much benefit.

The agricultural use of drip irrigation has however, been started in Israel. It is presumed that the method was devised in that country when a passer-by had noticed that some trees were growing more vigorously than those which were slightly away and the noticeable feature was that, there was a leaky faucet which was discharging water drop by drop through the leak and close to the vigorously growing trees. Thus, whether, discharge of water in drops had been a cause for more vigorous growth of the trees shook the attention of the passer-by, who pointed it out to the irrigation personnels.

The drip irrigation method that is used at present is an improved form which has been much simplified, made more useful and economic over the years. It has a *filter station* and a control panel with a network of main, sub - and lateral pipelines with *emission points* (= *wet points*) spaced at desired distance along their lengths. The emilters (= drippers) are specially made water issuing device, which discharge irrigation water drop by drop on the ground. The whole set up is mechanically regulated, automatically controlled and may even be computerized to ensure that measured and pre–determined amount of water is released at slow but regulated rate for desired duration.

The water is allowed to flow through the network of pipelines at pressure which is adjusted as desired. The drippers are attached at suitable points in the pipelines and the water conveyed through the pipelines is emitted from the drippers. The drippers are though distributed assymetrically in the soil, their placement is done close to the water–absorbing roots (= feeder roots) of the concerned plants. Thus, the soil closely around such roots is just moistened and does not become soggy, while other parts of the soil remain dry. The dripped water moves to the root zones due to the combined force of gravity and capillary action. As the drop by drop discharge of water is continuous, the withdrawal of moisture by the roots is replenished immediately and hence, they do not suffer from moisture stress.

Installation of the Unit

The drip irrigation unit is set up with the following components from the source of water supply to the delivery of water on the soil surface:

(i) Control Head

This is installed at the source of water and consists of a control-valve to regulate pressure of water and its flow rate, a volumetric water apportionner or timer and also a filtration sub–unit for filtration of the water against possible impurities.

Figure 10: Control Valves of different Types.

(ii) Main Distribution Lines

These pipelines are built up of plastic materials. The material should be strong, durable and the pipelines made with that should be of light weight, non–conducting, can be joined, shifted or modified easily, should have smooth inner–walls, withstand high degree of water pressure and non–reactive or non–corrosive to fertilizer salts or other agro–chemicals. PVC pipelines may serve the purpose. The length and the diameter of the pipelines very according to the rate of discharge of water as is actually desired and the area to be irrigated. The main pipelines are connected to the sub-pipelines and the latter to the lateral lines.

(iii) Sub– and Lateral Distribution Lines

These pipelines are made up of similar plastic material as with the main distribution lines but of smaller diameter (usually of 12–16 mm) than the main distribution lines. These are laid parallel to one another and attached to the main distribution lines with *connectors* (Figure 11).

Figure 11: Connectors

(iv) Drippers or Emitters

These parts are attached to the lateral lines which make water to trickle down for irrigation in drips. The drippers are mounted on the lateral lines known as mounted drippers (Figure 12). These are made up of strong and durable plastic material, should be leak proof and non corrosive to agro chemicals. However, the drippers may also be built–in (Figure 13) inside the lateral lines and this is done during their manufacturing. Irrigation water is made to trickle down from the drippers at a very slow rate, which usually ranges between 2–8 litres per hour. The rate of discharge can be adjusted by rotating the drippers with fingers so as to adjust the emitting holes.

Figure 12: Mounted Drippers

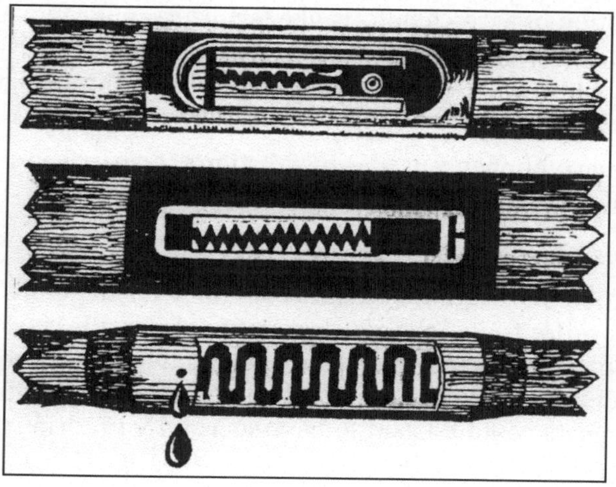

Figure 13: Built–in Drippers

Various types of drippers are in fact, available. Among these, of which the "Button Drippers" having water output capacity of 2, 4 or 8 litres per hour and which can be fixed on 12, 20 or 25 mm pipelines for longer line extension are usually preferred for irrigating in orchards. The "Pressure Compensated Drippers" have flow control regulators in them and ensure uniform flow of water. These are effective in sloppy land or in areas with low water pressure. For underground installation, "Woodpecker Drippers" are designed, which are buried under the soil and thus, are protected against damage by wood peckers, squirrels or birds.

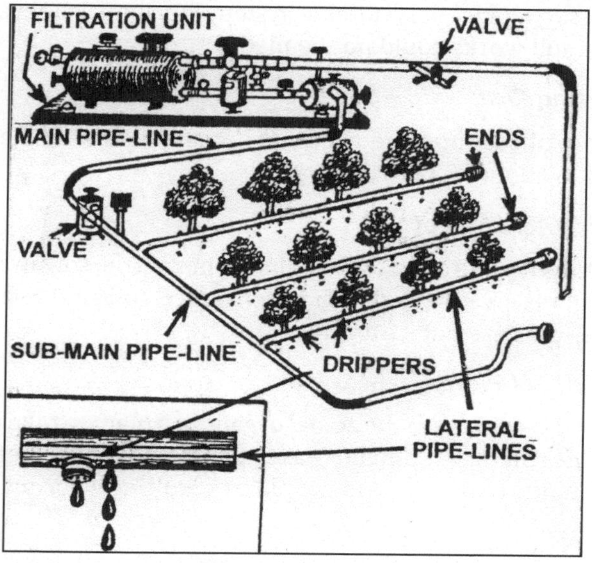

Figure 14: Drip Irrigation in Young Orchard

Where solution of agro-chemicals like fertilizers, nutrient salts, weedicides, plant protection chemicals *etc.* are desired to be applied along with irrigation water, extra accessories like tank to hold the solution, fertilizer pump (Figure 14) to inject the solutions in the pipelines *etc.* become necessary.

The different components used in installation of drip irrigation unit have been shown in the following figures (Figures 10–14).

(a) Advantages

Drip irrigation offers great potential especially for the crops that are grown in rows. Therefore, for fruit crops which are usually plants in rows, *e.g.*, pineapple, banana, grape, papaya *etc.*, irrigation by this method is of great help. However, for other systems of planting also, drip irrigation is proved to be efficient and at least the following merits could be said to be associated with drip irrigation in fruit plantations.

(i) Economy of Water

There is great saving of irrigation water by this method and it is estimated that in general, the method makes a saving of upto 70 per cent of water over the conventional surface irrigation method. The reasons for the saving are low discharge which prevents loss of water due to run off, preventing loss of water as results due to soaking of unwanted parts of soil, very low evaporational loss, control of weed growth *etc.*

(ii) Convenience in Irregular Terrain

To irrigate by drip method, levelling of land, making rings, channels, furrows *etc.* are not necessary. Hence, it is suitable in slopy land and in uneven physiography where the above soil works could not be made.

(iii) Low Operating Cost

The operational cost and the need for orchard operations and interculture may be reduced to about 50 per cent.

(iv) Unhampered Aeration of Soil

As the water moves through the capillaries in the soil, soil air is not driven out and soil macro–pores are not filled with water. Hence, soil air is not much disturbed and thereby, the roots are not suffocated.

(v) Better Growth of Plants

The feeder roots which are the actual organs in a plant to take up moisture and nutrients from the soil remify and increase in number by drip method. As a result, uptake of these inputs is increased in the plants resulting vigorous growth of them.

(vi) Weed Control

As very small part of the soil is moistened and remaining parts remain dry, germination of weed seeds and also growth of them are arrested in the drier parts of the soil.

(vii) Increased Soil Temperature

Soil heat is much conserved in this method as part of the soil remains dry. As the soil temperature is not lowered, the roots of many plants which are sensitive to low temperature are safeguarded.

(viii) Early Maturity

As the soil heat is conserved, early maturation takes place in grape vine and some other fruit crops in which cases, higher soil temperature favours early maturity of the fruits.

(ix) Irrigation under Polythene Mulch

Mulching the soil with polythene or other plastic mulches creates great difficulty to irrigate by surface application of water. In such situations, irrigation may be given by drip method through holes made in the mulch cover.

(x) Application of Fertilizers and Agro-chemicals

One great advantage of drip irrigation lies in the fact that the solutions of fertilizers, plant nutrient salts, weedicides, plant protection chemicals *etc.* may be applied in desired amount through the drippers. By such type of application, much less amount of the inputs are needed, their placement is done at exact locations arid the extra cost that is involved in conventional application of them is avoided.

(xi) Disease Control

As the soil moisture and chemical additive levels are closely controlled, disease incidence is lesser in this method. Also that the disease infection in healthy trees as may results to transfer of the diseased leaves, twigs, fruits *etc.* from one part of the orchard to the other part by the irrigation water in case of other surface irrigation methods (*vide* 5–b) is avoided in drip irrigation.

(xii) High Yield

By ensuring optimum growth of the crop plants, drip irrigation proves to be of great help in raising their yield performance. The yield may increase from 20–50 per cent although increase to more than 200 per cent has even been claimed in some agricultural crops.

(b) Disadvantages

Despite the above merits, drip irrigation method has a number of drawbacks also, among which the following are mentionable:

(i) Clogging of Pores in the Drippers

Clogging of the dripper orifice is perhaps the most serious problem. When very low rate of water flow, *i.e.*, 1–8 litres per hour is to be dripped, it becomes necessary to make narrowing of the passage and orifice of the drippers. This is usually made less than 1 mm in diameter such that the average diameter becomes 0.5 mm. Making so narrow outlet of a dripper very often causes its clogging and this results in failure of uniform dripping of water. This becomes more problematic, if solutions of fertilizers or other agro-chemicals are used in drip irrigation.

Washing the pipelines and the pores of the drippers by acid is however, done by many users to clear off the passage. For this purpose, dilute sulphuric acid is usually injected into the unit to wipe out the clogged materials. But this may be dangerous because pouring acid on soil changes its *p*H and may also cause health hazard. It is however, pointed out in this connection to the experiments done by Schwanki and Pichard in the nineties in California, USA. They observed that low concentrations of the chemicals, ESI–50 which is a phosphonate compound and is used as a scale inhibitor in the factories, if injected into the pipelines of the drip irrigation unit made vary good clearance of the clogging materials, than the commercial product, N–pHURIC, which is generally used for the purpose.

(ii) Filtration of the Irrigation Water

Ordinary water usually contains many impurities. These may be physical, *e.g.*, dirt and dust particulates, lime *etc.* or chemical, *e.g.*, salts of calcium, sodium, magnesium, iron, chlorine *etc*, or biological such as algal, fungal and bacterial growth, protzona *etc.* If such type of water is directly passed through the pipelines, these become narrowed by their deposits, the drippers are plugged and chemical reaction takes place which may be corrosive to the unit. Therefore, it becomes necessary to filter the irrigation water before it is passed through the pipelines and accordingly, this makes an extra expenditure in drip irrigation.

(iii) Adjustment of Pipelines and Drippers

The pipelines and drippers should be laid on ground with very careful adjustments. Shifting their positions during progress of growth of the fruit trees becomes necessary in an orchard when the method is used from early life of the orchard and evidently, this involves labour and regular watching.

(iv) High Initial Cost

Despite the fact that drip irrigation has a number of merits, high initial cost is involved in installation of the whole set–up. Hence, it is difficult to be afforded by many farmers and orchardists in India who belong to the low income group and do not have adequate area of land for cropping.

9. Perforated Pitcher (Surface drip) Method

This is crude type of drip irrigation method. To make drip irrigation by this method (Figure 15), fairly large sized earthen pitches should be taken. A small hole should be made at the bottom of each pitcher. Well burnt, 3 strong bricks should be placed in a triangular manner or 4 bricks placed in a square on the ground at the location where dripping is to be done. Each pitcher is carefully placed on the 3 or 4 bricks in such a manner that the hole at its bottom remains free and is not blocked.

The pitcher should be filled with water and the water will drip down on soil through the hole of each pitcher. These should be filled with water every day or when they become empty. Filling of water may be done by conveying the water through flexible pipes. To avoid drinking of the water in the pitchers by birds especially if fertilizers or agro-chemicals are added to it, the pitchers may be covered with earthen lids having a small hole in each or any suitable material. Instead of earthen pitchers, old tins may also serve the purpose making holes at their bottom. The method has been much popular to irrigate orchards in West Bengal.

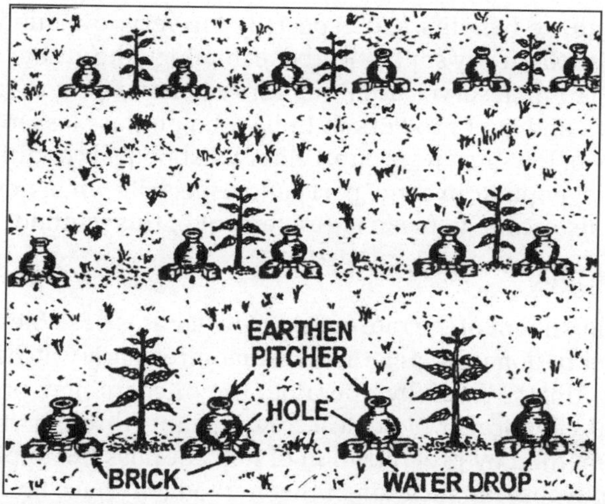

Figure 15: Drip Irrigation by Performed Pitchers in Young Orchard (*Courtesy*: Principles and Methods of Orchard Establishment by Dr. B.C. Mazumdar, 2003)

(a) Advantages

The following advantages are undoubtedly derived by this method:

(i) Very Low Cost in Installation

As compared to normal drip irrigation method consisting of different parts and accessories, the cost of installation of this method is extremely less.

(ii) Low Recurring Cost

Unlike drip method, recurring cost is also extremely less in this case. If carefully handled, the earthen pitchers or tins could be used for long time without causing breakage of them.

(iii) Simple to Operate

Any accessory or sophisticated technology is not involved in this method unlike that of drip irrigation and hence, an ordinary farmer or an orchardist can use it freely.

(iv) Use of Unfiltered Water

Unlike filtration of the irrigation water where clearing of the possible impurities becomes necessary before it is passed through the pipelines in drip irrigation, in this method, unfiltered and ordinary water may be used for the purpose.

(v) Application of Fertilizers and Agro-Chemicals

Like drip irrigation, fertilizers, nutrient salts or other agro chemicals may also be applied in required amount by dissolving the same in the water of the pitchers.

Due to the above benefits and especially for very low initial and recurring costs, simplicity without any sophisticated technology, the author is of the opinion that pitcher irrigation deserves attention for recommending to the ordinary and marginal farmers and fruit growers in India who possess lesser area of land for cropping and are of low-income group. Under West Bengal conditions, the method has become much popular in some parts and especially when new orchards area set up. There is also scope to make improvement of this method.

(b) Disadvantages

Accurate adjustment of the rate of flow of water as is possible in drip irrigation could not however, be done in this method and unless regularly watched, over- dripping may also result. Unless carefully operated, the pitchers may break. Prohibiting entry of physical, chemical or biological contaminants through irrigation water as is done in drip method by pre–filtration of it is not possible in pitcher irrigation. For large scale purpose, application of the method may have limitation.

10. Pipe Method

Irrigation water may also be delivered on a cropped plot conveying the water through flexible pipes. However, this is possible when water is pumped out into the pipeline or is sent into it through gravitational force. Hose–pipes were formerly used for the purpose but nowadays, pipes made of plastic and sometimes rubber tubings are more commonly used, because of lesser weight, more flexibility and also lesser cost. The diameter of the pipes used may be 3 to 12 cm, in accordance with the amount water that is to be delivered and the rate of discharge and also the pressure.

The water discharging end of such a plastic made pipe is sometimes fitted with a detachable faucet having 2–4 openings (Figure 16). This is attached while discharging water with the idea that water is delivered at different directions instead of at one place and the water pressure is also reduced. Thereby, heavy watering at one place with high pressure and causing depressed pocket as may happen there if the faucet is not joined is avoided.

Thin gauged polythene made tubings available in rolls (ribbons) are also much used now a days for very low cost. However, durability of these ribbons is very less and can be temporarily used only.

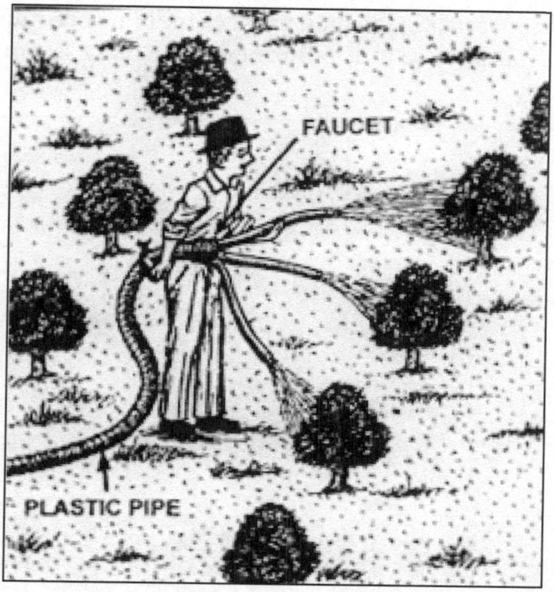

Figure 16: Irrigation by Discharging Water through Plastic Pipe Fitted with Detachable Outlet of Several Openings.

(a) Advantages

Plastic tubings are very convenient to use. Irrigation by this method can be done at desired level and when plot has been irrigated, it can be shifted to the other plot. The method has great suitability to irrigate inter crops in the orchard.

(b) Disadvantages

The durability of these tubings is not high. There is chance of damage at any part of the tubing easily and thereby, leakage of water may result. Damage by squirrels is also not ruled out.

11. Soil Sloping (= Contour) Method

This is a method of collecting rain water rather than artificial irrigation. The method consists of raising the inter-row spaces with soil such that the level of the rows of fruit trees remains lower than the level of the space between the rows. To raise the level of the inter-rows, soil is heaped at the centre to different heights as desired, which usually ranges from 20–40 cm above the ground level. On doing so, the centre of the inter-row spacing remains at the higher level and slopiness is maintained on both sides of the raised part. Evidently, when there is rain, water that falls on the inter–row spaces flows down along the slopes on both sides and is accumulated in the rows of the fruit trees (Figure 17). The water stand each row of fruit trees as the level of it remains lower than the both sides. The standing water is gradually percolated to lower depth.

The method is observable in some parts of India where rainfall is scanty and the soil is porous in nature and is usually done in the rainy season only rather than throughout the year.

Figure 17: Soil Slopping Method

(a) Advantages

This method of efficient utilization of rain water, rather *rain water harvest* technique is found to be much help to irrigate fruit trees in areas where rainfall is very low and soil is of loose texture. As the rain water falling on the orchard is diverted directly to the fruit trees in the rows, they get ample amount of water. The water is also conserved in the soil over a longer period.

(b) Disadvantages

The method has a great risk in that, there is erosion of soil from the inter-row spaces. Due to slopiness on both sides from the middle-part of the inter-row spaces, soil is eroded and is collected in the rows of the fruit trees. This results in burying part of the tree trunks in the ground, which is harmful as this leads to infection and rotting of the trunks or their collars. Much labour is also involved in raising the level of the inter row spaces. If there is any excess rainfall during the season, much amount of water is accumulated in the rows of the fruit trees, resulting water logging. In such situation, the excess water needs to be drained out to save the trees.

Chapter 9

Irrigation System: Sub-surface Irrigation

This is a system of irrigation in which, irrigation water is discharged in the sub–soil layer *i.e.*, at a depth usually of 20 to 40 cm from the ground level. The idea of such type of irrigation is to deliver the water as closely as possible to the water absorbing (= feeder) roots so that those can absorb water quickly, which may be difficult or takes time especially in heavy clay soil when surface irrigation is given. There are a number of methods under this system which have been discussed in the following but it should be noted that many of these methods are rarely practised because of practical difficulty and especially as surface-drip and sprinkler methods prove superiority over those methods.

(1) Straight Trench Method

This method of irrigation could be considered to be a modification of the furrow method as described earlier under the Chapter 3 and the differences with the furrow method are as follows :

(i) The furrows are made deeper in this method, *i.e.*, 20 to 40 cm depth so that they resemble trenches rather than furrows.

(ii) One end of the trench is blocked in order that water is held in them, unlike the furrow method where the water on flowing through the furrows is given off to the drainage channel.

(iii) The number of trenches that are made are lesser in number than the number of furrows that are opened in the same area of land.

The number of trenches vary but at least, two trenches are dug one on each side of each row of fruit trees and at such position that it remains just below or slightly away from the sidewise spread (= canopy of growth) of the trees. Thus, the trenches run parallel to each row of trees and just below or at a little distance from the skirt of the trees but in one direction only (Figure 18). Besides these two, more trenches may be dug, if desired by the orchardist and in the inter-row spaces.

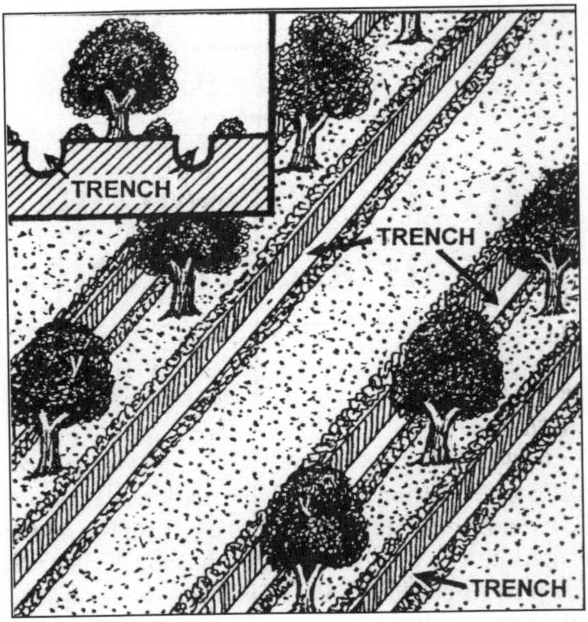

Figure 18: Straight Trench Method.

(a) Advantages

Where percolation of water has difficulty as in heavy clayee, soil, the trench method may be advantageous over furrow method as the irrigation water is directly delivered to the sub-soil layer in this method. Thus, the water absorbing feeder roots get the water directly and hence, take quick growth. During rain, the water also remains stored in the trench. For deep rooted fruit trees, this method may be of utility.

(b) Disadvantages

In most cases, the method does not gain support because many thick, anchorage roots and the thin feeder roots of fruit trees are *cut off* while making the trenches. Stagnation of water in the trenches may also result in this method which sometimes brings about suffocation to the roots. Besides, cost is also high in this method for much spade work and manual labour necessary to dig the trenches. Maintaining the trenches by regular removal of soil which deposits there also involves an expenditure.

(2) Cross Trench Method

The method is similar to straight trench method with the addition that the trenches are dug crosswise, *i.e.*, in both directions. Thus, the trenches run in both directions for squares or rectangles (Figure 19).

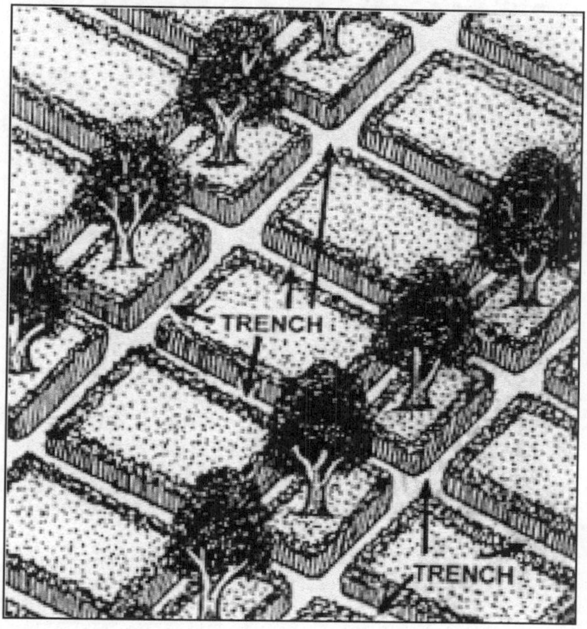

Figure 19: Cross Trench Method.

(a) Advantages

As trenches are made in both directions, greater parts of the roots are irrigated than the straight trench method. Much greater amount of water is also held and conserved in the orchard. Besides, the trenches offer protection to the trees as well against possible entry of animals.

(b) Disadvantages

Maintaining so many trenches on the orchard land is hardly justifiable. These may cut off many roots and hence, the fruit trees in the orchard may be devitalised or may even fall off. There is also likelihood of water stagnation, which may hinder aeration of the roots.

(3) Circular Trench Method

In this method (Figure 20), a circular (or may be a square) type or trench is dug on the soil separately around each tree. The trench is dug upto the similar depth as in the above methods, *i.e.*, 20–40 cm and is positioned just below the peripheral spread of the tree or slightly away from that. In some cases however, trenches of much smaller diameter are also made.

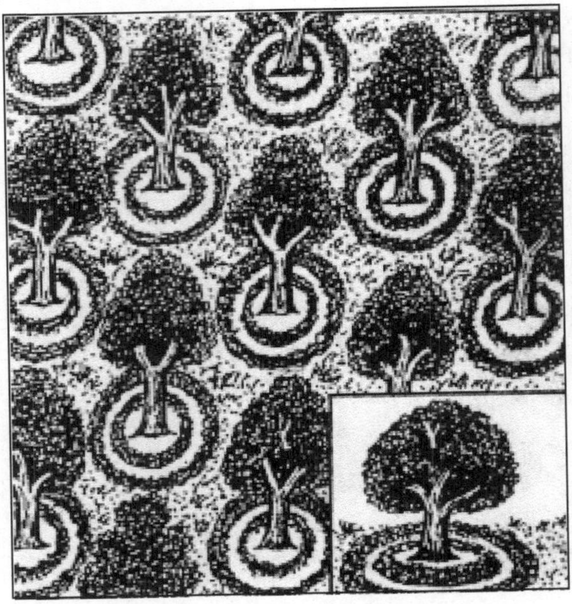

Figure 20: Circular Trench Method.

(a) Advantages

The method is usually adopted for young trees to serve as protective measure against grazing of them by animals and in areas where regular watch and ward are not possible. Another purpose is to hold water in the rainy season and to conserve it where artificial irrigation is not possible to be done. These advantages are derived by this method and if artificial irrigation is done, the trenches may even be inter-connected with one another by making sub-channels as in the ring or basin method.

(*N.B.* It should be noted that the soil sloping method as described in the earlier Chapter on Surface Irrigation System and the straight trench, cross trench and the circular trench methods as described above in this Chapter are adopted to make more utilization of the rain water and are not much related to irrigation by artificial means).

(b) Disadvantages

Except to drive away entry of grazing animals, the method has little suitability for the same disadvantages as mentioned in straight trench or cross trench method.

(4) Perforated Pipeline Method

In this method, perforated pipelines are placed at the sub-soil upto about 40 cm depth and are covered with soil. Thus, they remain embedded underground.

The pipelines are made up of iron, aluminium or suitable alloy, the diameter of the perforations may be 2–4 mm and the number of perforations per square centimetre may be 4 to 9 or so. One end of each of these perforated pipelines which remains under the ground is sealed while the other end is connected to the main pipelines which is not perforated. Irrigation water is passed through the main pipelines at desired pressure. This goes to the perforated lines and water is ultimately discharged in the sub–soil through the perforations and thus, made it soaked.

(a) Advantages

Apparently, the method seems to offer the following advantages:

(i) Irrigation water is delivered near to or directly to the water adsorbing roots of the trees. Therefore, particularly for deep rooted fruit trees, the method is a good contrivance to supply water quickly and directly to the soil close to the water–absorbing roots, which may not take place so easily in surface irrigation methods if the percolation rate in soil is low.

(ii) Evaporational loss of water is avoided as the water is directly supplied to the underground.

(iii) The soil surface remains dry, for which emergence and growth of weeds are greatly controlled.

(iv) In irregular terrain where surface irrigation methods are difficult, this method has greater suitability.

(v) The difficulty in surface irrigation as may result where plastic or inorganic mulches have been used may be avoided in this method as the water is delivered underground and not on the surface.

(vi) The solutions of fertilizers, plant nutrient salts or soil disinfecting chemicals may be supplied directly close to the root zones and in the sub-soil.

(vii) Another important advantage of the method lies in the fact that raising temperature of the sub-soil may be done by supplying hot water through the pipelines. Accordingly, the method proves benefit in temperate zone and also in sub-tropical zone where sub-soil temperature sometimes goes low or very low due to snow, frost or sudden fall of atmosphere temperature.

(b) Disadvantages

For the advantages stated above, the method had been followed decades back in some parts of the western countries and even in few parts in India also to irrigate some crops notably, grape, strawberry, some temperate fruit trees and shrubs and perhaps in betel vine, which is however, not a fruit crop. The method was followed for ornamental plants as well, grown in parks and gardens, might be to restore the beauty as would be hampered by installation of surface irrigation methods. At present, the method is rarely practised under commercial scale for the following demerits and especially after introduction of surface - and sub-surface drip irrigation methods.

(i) The discharge of water through the perforations is ordinarily difficult to control, unless the flow of water is properly regulated at the source. As the discharge at the delivery point is not under control, the method may result in over–watering in the sub-soil. This may be greatly harmful as the water-table may be raised, drainage of free water may be hampered and suffocation of roots may take place.

(ii) The hazards of breakage, distortion, flattening, corrosion and also rusting of the pipes if made up of iron are also there in the method.

(iii) Gradual narrowing of the inside of the pipes affecting easy flow of water as a result of deposition of calcium or other salts is not ruled out, if unfiltered water is used.

(iv) Clogging of the pores in the pipelines very often results in this method and cleaning of them without removal from underground is difficult.

(v) Evidently, the cost of installation is very high and this high cost perhaps is not compensatory to the advantages that are derived by this method.

(5) Trench Drip Irrigation Method

This is the drip irrigation method as described in the Chapter 4 with the only difference that the lateral pipelines (= drip-lines) along with the drippers are placed in trenches. Thus, to install the method, trenches are at first dug in the same way as described above under the straight trench method in this Chapter. When the trenches have been dug to the desired depth, the lateral lines containing the drippers are laid down inside the trenches. The water is dropped inside the trenches from the drippers and thus, dripping takes place in the lower layer of soil, in other words, close to the feeder roots. Therefore, unlike the dripping on surface soil as is done in drip, *i.e.*, surface-drip method, dripping in the sub-soil is brought about by this method.

(a) Advantages

(i) In soil of high clay content where rate of percolation is low, this method offers to have advantage to make the water reach the soil mass near the root zone directly. In such type of soil or where an impermeable layer exists in the soil near the surface, surface drip irrigation does not prove to be of benefit as the water flowing in drops has difficulty to percolate and it actually evaporates before percolation. Therefore, dripping in trenches is more efficient in such cases.

(ii) In case of deep rooted fruit trees, the method makes easy, quick and direct discharge of irrigation water near the root zone.

(b) Disadvantages

A number of feeder roots may be damaged or cut off in making the trenches besides the fact that an extra expenditure is involved in digging them.

(6) Buried (= Concealed) Drip Irrigation Method

In this drip irrigation method, the pipelines (= drip lines) and the drippers are buried under the ground either completely or partly. To install the trench drip irrigation method as described above is implemented, making the depth of the trenches from 20 to 40 cm as desired. The drip lines are laid in the trenches and then the trenches are filled with soil so as to embed the drip lines. The drippers (= emitters) may be buried when complete concealment is done or projected above the ground when partial concealment is desired.

When the drippers are embedded under the ground, short tubes of smaller diameter are connected to the drippers which open upward above the soil surface. The tubes may be held in suitable position with the help of stakes or suitable supports so that dripping takes place on desired spots from the tubes (Figure 21).

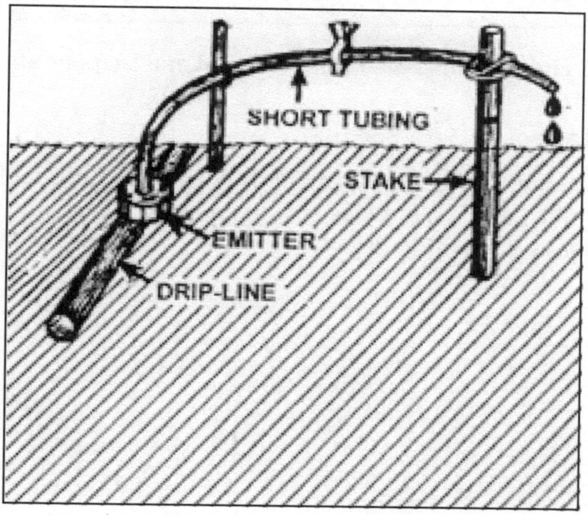

Figure 21: Drip Line and Dripper Concealed Underground and Dripping Done on Surface though Tubes

In case it is desired to maintain the drippers above the ground, the short tubes of smaller diameter are connected to the embedded drip lines but the drippers remain outside. That is, one end of a tube is connected to the drip line and the other end to the dripper. The dripper may be held in suitable position above the ground with a stake and water is discharged from the dripper in drops (Figure 22).

(a) Advantages

The same advantages as are derived in trench drip method are obtained in this method also. As the drip–lines are concealed, they are protected against damage by wild animals. Besides, weed growth is usually much lesser in this method.

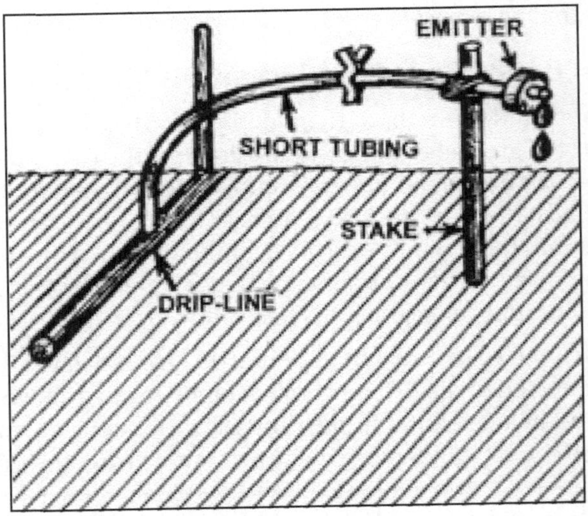

Figure 22: Drip-line Concealed Underground and Dripper above Ground

(b) Disadvantages

Due to concealment, damage if any in the drip-line is difficult to be detected from above. In some soil, *e.g.,* in black cotton soil which is easily hardened, lifting of the drip-lines may have difficulty.

Chapter 10

Irrigation System: Overhead Irrigation

In this system of irrigation, fine droplets of water are sprinkled above the ground level in the orchard. Hence, the system is also known as *aerial,* or more commonly as *sprinkler system of irrigation.* The system was innovated from the principle, "the best source of irrigation for crop plants is nature in the form of rains". Thus, sprinkler irrigation consists of pouring water like rain. The system is highly practised in the U.S.A. and many other countries. In India, it is now considerably practised in parts of Haryana and Karnataka and to some extent in Gujarat and in West Bengal in the decorative gardens.

The system essentially consists of a water–pump which is connected to the main-pipeline. The main pipeline is coupled with the sub-main pipelines and the sub-main pipelines with the lateral pipelines. The lateral pipelines are connected to the vertically erected posts and the posts are finally fitted with sprinkler. Irrigation water is supplied to the main pipeline at high pressure. As a result, the water is conveyed to the sub-main lines, then the lateral lines and as shown in figure 28, the droplets of water are finally thrown out by the sprinkler nozzles. The rate of discharge of droplets as well as the size of droplets can be adjusted and when discharged in extremely fine from, it is known as atomization and when in still fine from, the term *mist* is used as it is like a mist.

The sprinkling of water in the orchards or other plantations of fruit crops is generally done by one of the two days, which are *(i) over the tree sprinkling* and *(ii) under the tree sprinkling.* In the former case, water is sprinkled over the branches including foliage parts of the trees and is also known as *complete coverage spray* of water. But in the latter case, water is sprinkled on the ground only, *i.e.,* beneath the crowns of the trees and not on the foliage parts. The former is no doubt, more commonly practised. However, many fruit growers even prefer to make a combination of both, especially when intercrops, cover crops *etc.* are grown in the orchard (*vide* Part 2) or for some other purpose.

Installation of the Method

As in case drip-irrigation, in sprinkler irrigation also, a number of accessories are involved to set up the method and the principal ones may be listed as follows:

(i) Pumping Set

For the purpose of spraying water through the sprinkler nozzles, water should be pushed to the pipelines at high pressure, otherwise sprinkling would not be possible. The pressure at the pipelines may be 70–120 psi, depending on the length and diameter of the pipelines, distance of the pipelines, height of the sprinkling posts (= riser posts), rate of discharge of water, distance of water desired to be thrown out from the sprinklers *etc.* Accordingly, pumping set at the water source becomes necessary to be installed for introducing water in the main pipelines. However, if the water source is located at a great height from the area to be sprinkled as in case of a dam, reservoir in hills *etc.*, pumping set is not necessary as the vertical pressure of water may be enough for the purpose although it depends on the height of the water source.

(ii) Filtration

While in drip irrigation, filtration of the water before it is sent to the pipelines is a must, in sprinkler irrigation, filtration is or is not done. But filtration of the water is definitely better because unfiltered water containing impure particles may clog the sprinkler nozzles and thereby, cause blockade in sprinkling. Besides, the pipelines may also be corroded due to the presence of salts in the water.

(iii) Main Distribution Lines

The pumped out or vertically descending water as the case may be is on filtration taken up by the main distributing pipeline. The water received by the main line is flown through this at high pressure.

(iv) Sub-main Distribution Lines

Where the area to be irrigated is very large, it becomes necessary to provide sub-main distributing pipelines. The sub-main lines are of lesser diameter than the main lines and these are connected to the latter by suitable coupling arrangement (Figure 23). These lines receive the water from the main lines.

(v) Lateral Distribution Lines

These pipelines are still narrower than the sub-main distributing lines and are connected to those with couples. For greater efficiency in distribution of water, *valves* are sometimes provided inside the lateral lines and near to the couplers.

The water from the sub-main lines is delivered to these lines and obviously, the pressure is increased.

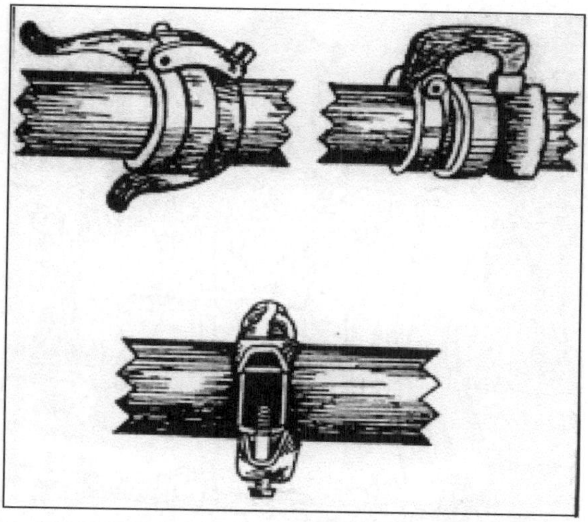

Figure 23: Coupling of Pipelines.

The main, sub-main, if any and the lateral lines may be installed on the surface of the soil, or may be completely or partly buried, if their concealment is desired. The depth at which the lines are buried may be 30–60 cm depending on the soil, tillage operations *etc.*

It should be remembered that unlike drip irrigation which operates at a very low pressure of water, sprinkler irrigation functions at *very high pressure* only. Such pressure is usually not withstood by plastic made pipes that are used in the former case. Hence, the main, sub – and the lateral pipelines used in sprinkler irrigation are built up of hard metal or alloy. The pipelines made up of lightweight galvanized steel or aluminium or aluminium alloy are however, more commonly used. The aluminium pipes are rust resistant and of much lighter weight than those which are of galvanized steel and hence, are portable from one place to another (A 1–metre long aluminium pipe of 75, 100 and 125 mm diameter usually weighs 0.9, 1.2 and 1.7 kg while a lightweight galvanized steel pipe of similar length and diameter may weigh 2, 3 and 3.5 kg respectively). On the other hand, the galvanized pipes are less costly, stronger, more durable and superior in respect of denting and bursting capacity.

(vi) Riser Posts

These are vertically posted pipes, the basal ends of which are connected to the lateral lines with the help of couplers and the upper ends with the sprinkler nozzles. Accordingly, these receive water from the lateral lines. The diameter of the riser posts becomes lesser than the lateral lines and hence, the water pressure is further increased. The length of the riser posts vary and may be from 0.8 to 2.5

metres. However, with couplers and valves, the height of the riser pipes may be made increased or decreased without stoppage of the pump (Figure 24).

Figure 24: Raising Height of Riser Post.

(vii) Sprinkler Nozzles

These are the ultimate components of the unit and are fitted at the upper ends of the riser posts. The water taken up by the nozzles from the lateral lines is finally sprinkled outside at high pressure.

Various types of sprinkler nozzles are available (Figure 25). These may be stationary *i.e.*, having no movement or of swivelling type, in which case, they themselves rotate either partly or fully around the axis and while rotating, they throw water droplets. The stationary type sprinklers are though less costly but the total cost is higher in their installation as several number of these are required for pointing towards different directions.

It is obvious that for the purpose of sprinkling water, very high pressure needs to be maintained at the sprinkler nozzles. The pressure may be 40 to 90 psi (= 3–6 kg per sq cm), depending on throw distance of water, rate of discharge of water and other factors.

The flow rate of water from the sprinklers is however, adjustable and may be regulated to release water droplets which may range between 4 and 200 litres per minute or more.

In addition to the above accessories, if it becomes necessary to supply fertilizer salts in solution, *fertilizer tank with injection arrangement* (Figure 26) is to be attached to the unit. These tanks are usually made up of steel or aluminium alloy with lids and coated with plastic material to protect against weather. The tanks of various capacities are available and that of the required capacity should be used.

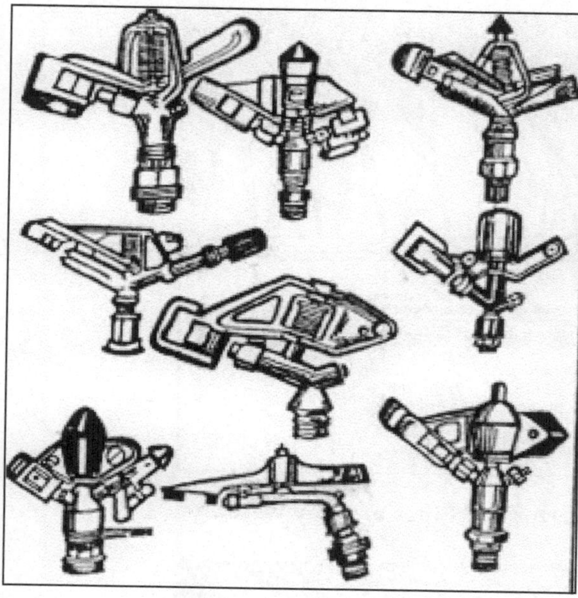

Figure 25: Sprinkler Nozzles of Different Types.

Figure 26: Tank to Inject Fertilizer Solution.

In the recent years, there has been a lot of technical improvement including computerization in the sprinkler irrigation units used throughout the world. These have enabled to make sprinkling or atomizing water at sufficiently long distance in measured amount, height and for required duration of hours. Where *intermittent sprinkling* rather than continuous sprinkling is desired, time switch needs to be attached to the unit. Evidently, the initial cost goes higher in the improved types.

However, simple types without much sophistication are also available and these have been improved without enhancing the initial cost. A low cost unit which is simple and consists of perforated aluminium pipes (Figure 27) has gained popularity now. Water is sprayed from these perforated pipes but at much low pressure.

Instead of installing any permanent type of sprinkler irrigation unit, even spraying water over the trees with an ordinary power sprayer or knapsack sprayer also produces some good effect when spraying is done over the foliage parts in hot weather at suitable intervals in tropical and sub-tropical situations. Fertilizers or plant nutrient salts in solution form may also be sprayed over the trees. It is a practice among the fruit growers in many countries to spray the trees a fortnight or a month before the harvest. This is known as "fruit sizing spray" as the fruit size is increased by this. Spraying with even ordinary water increases size of the fruits but perhaps lowers their keeping quality. As the fruits are harvested shortly after spray, fertilizers, plant nutrient salts or any other chemicals that may render harmful effect to the consumers should not be applied in any case.

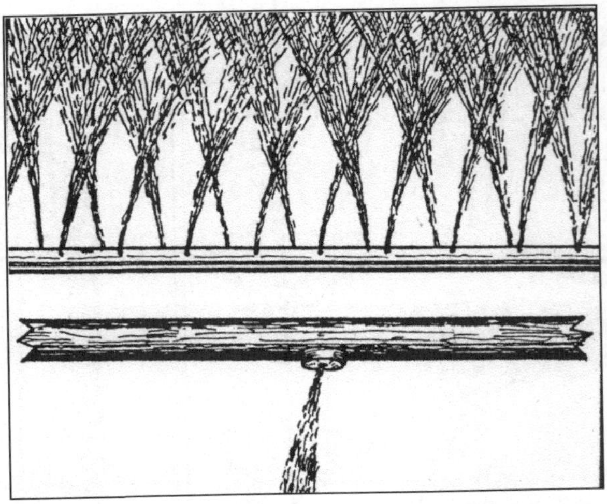

Figure 27: Perforated Pipes to Spray Water.

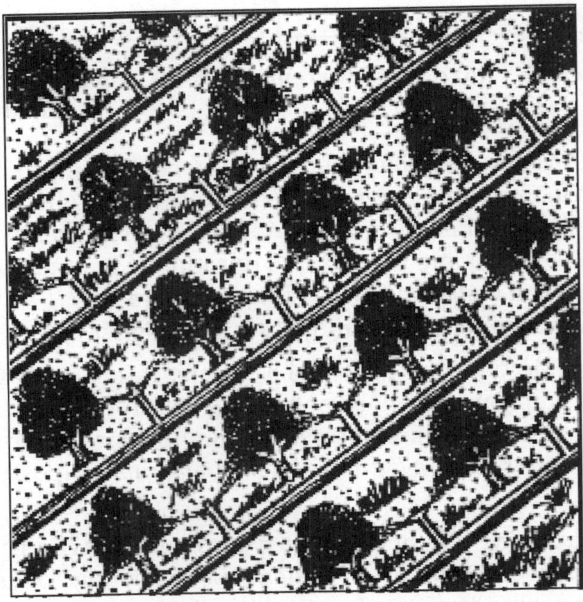

Figure 28: Water Droplets Sprinkled from the Nozzles in Sprinkler Irrigation Method.

(a) Advantages

Sprinkler irrigation is considered by many as the best of all methods of irrigation, although a section of irrigation technologists is of the opinion that drip irrigation as the best method. However, the definite advantages of the sprinkler method may be stated as follows:

(i) Washing of Foliage and Plant Parts

The trees are thoroughly bathed by spraying water over them. As a result, eggs, larvae and pupae of many insects and fungal spores if present in them are washed out. Thus, control of pests and diseases is done to a considerable extent without application of harmful pesticides. Besides, the dirt and dust which block the stomatal openings in the leaves are washed and as the openings remain free, transpiration rate is increased, eventuating higher uptake of water and minerals. As the leaves are washed, they become more efficient in respect of photosynthesis and respiration.

(ii) Lowering Air Temperature

The air temperature in the inside of the trees is lowered by sprinkling water and hence, is of much help to ward off solar injury which results in many trees by strong sunshine. The drying of terminal shoots, if results due to moisture stress is avoided by this method of irrigation.

(iii) Better Pollination in some Fruit Trees

As the relative humidity in the inside part of the trees is increased by spraying with water, pollination is also increased in some fruit trees. Custard apple is a good example of such type of fruit trees. In Egypt and also in northern India, it is noticed that most of the flowers in this tree which appear in summer months are not pollinated as the period of receptivity of the stigma is curtailed and hence, the flowers fall off. But as the relative humidity is increased in rainy season, pollination is also found to be increased. It has been experimentally verified that spraying water in the inside of the trees at suitable intervals in hot weather improves pollination to a great extent by increasing the relative humidity.

(iv) Frost Protection

In the recent times, protection of orchards against frost by "under-tree sprinkler irrigation" is gaining much prominence throughout the world. The heat is mainly radiated from the soil surface due to sprinkling of water and for this purpose, sprinklers are spaced at varying distances depending on severity of frost, type of fruit crops grown, discharge of water and other conditions.

(v) Convenience in Slopy and Uneven Land

Levelling of land, digging of channels *etc.*, are not required in this method and hence, like drip irrigation, the method is very suitable in hilly areas and in undulated topographic situation.

(vi) Application of Fertilizers, Plant Nutrient Salts and Plant Growth Regulants

The fertilizers, nutrient salts and plant growth regulators which are soluble in water may be applied easily to the desired concentrations while irrigating them. Especially the various micro-nutrients and growth regulators which are needed in very small amount are best applied to the trees by spraying solutions of them along with irrigation. In other methods of irrigation, this is however, not possible.

(vii) Economy of Water

Like drip irrigation, sprinkler irrigation also makes great saving of water, which may by 30–50 per cent over the surface irrigation methods. Besides, dispersal of water is also done very uniformly in this method.

The comparative performance of surface, sprinkler and drip irrigations as presented by Sivanappan (1988)* is stated in the following table.

Irrigation Efficiency of Various Systems

Irrigation Efficiency	*Percentage (in Methods)*		
	Surface	*Sprinkler*	*Drip*
Conveyance efficiency	60–70	100	100
Surface water moisture evaporation	60–70	70–80	90
Overall efficiency	30–35	40–60	75–80

The data indicate that sprinkler irrigation is about 50 per cent more efficient than the surface methods but efficiency of drip is 2–2.5 times greater, than that of surface and about 1.5–2 times more than sprinkler method.

(b) Disadvantages

Among the disadvantages of sprinkler irrigation, the following are mentionable:

(i) Ineffectiveness for many Plants

The sprinkling of water over the foliage parts of the trees does not have much effectiveness for those trees, the leaves of which are coated with waxy materials like cutin, suberin *etc*. The presence of these substance act as water proofing materials and accordingly, the water droplets falling on the leaves trickles down and fall on the ground instead of foliar absorption. Thus, the soil is though moistened, direct uptake of water by the leaves is obstructed.

(ii) Adverse Effect during Flowering Condition

Although there are some trees where pollination is favoured by high relative humidity as stated above, there are many other trees where the requirement is opposite. That is, creating high relative humidity inside the trees hampers activity of pollinizing insects in them and thus, pollination is much hampered. Besides, sprinkling water during flowering time may also wash away anthers, pollen grains and the stigmatic sap and thereby, prevents pollination. There may be heavy drop of flowers as well by spraying water during flowering time, particularly if preceded by a spell of dryness and the same effect may be for newly set fruits.

* Sivanappan, R.K. (1988), Irrigation–Times for new strategies. *The Hindu Survey of Indian Agriculture.*

(iii) Hampering Free Movement of Air

High relative humidity created by spraying water may affect the concentration of oxygen and carbon dioxide gases in the atmosphere of the orchard and this may be to a greater extent by spraying water in the inside part of the trees. Accordingly, the gaseous exchange rate of the trees may be hampered.

(iv) Harbouring of Pests and Diseases

High relative humidity created in the interior part of the trees by drenching water may provide congenial atmosphere for the spread of many pests and diseases.

(v) Water Stress in Roots

Inadequacy of sprinkled water may fail to provide optimum moisture that is needed by the roots of the trees. As a result, growth of lateral roots and root hairs may be hampered.

(vi) Unutilization of Sprinkled Water

In some plants, *e.g.*, tomato, maize *etc.*, it has been reported that the water taken up by the leaves travel in a *negative direction* inside the plant body and ultimately, the water is diffused out into the soil from the roots. Such a phenomenon usually results when the *diffusion pressure gradients* favour movement of water in the direction of roots. The condition if happens in fruit crops would prevent utilization of the water that would be applied as foliar spray.

(vii) Clogging of Pores of the Sprinkler Nozzles

As in case of drip irrigation where clogging may take place in the drippers, in sprinkler irrigation also, clogging of the pores of the sprinkler nozzles may result particularly when unfiltered water is introduced into the pipelines.

(viii) High Initial Cost

High initial cost is no doubt, a handicap for large scale use of this method.

Chapter 11

Precautions of Irrigation

Irrigation to a crop plant does not mean delivering water to it only but other considerations also are there. An irrigation engineer knows from where and how irrigation water should be brought and convey it near to the area where crops are grown. But to an agricultural and a horticultural technologist, many other considerations are necessary to be made if the irrigation water is to be made use of real benefit to the crop plants. As a matter of fact, even if the ideal system and the method of irrigation have been adopted in an orchard, the irrigation may not render any advantage to the crop and instead, is liable to cause serious and even irreparable damage to the soil and great harm to the crop if adequate precautions are not taken in its operation. It should be kept in mind that the orchardist should have a thorough knowledge of the soil and requirement of water by the fruit crops at the time when irrigation is to be done. On having this knowledge, he should take into account a number of considerations before, during and after the irrigation, if the benefit of this valuable input of the crop is desired. In fact, there are a large number of factors that are necessary to be considered and the major ones have been presented in the following.

1. Drainage

The term, drainage when applied in agricultural, horticultural or forest land refers to removal of the excess water from the site which has received water either naturally or artificially after the required amount of water is taken up by the crop if any is grown there and at a given point of time. Thus, it is a process to get rid of the unwanted water at any site.

Beyond doubt, any agriculturist or horticulturist hhaving some experience of crop husbandry is well aware of the fact that *drainage is an aspect which is no less important than irrigation* in successful culture of a crop plant. Again, the progressive fruit growers know that even if other hazards are tolerated by a fruit crop to some extent but in no case, if drainage is hampered. As a matter of fact, *drainage has a vital role in fruit culture* and the lack of it seriously stands in its way.

The excess water is drained out from the field in many ways. It should be noted that in every land, some part of the water is always drained out by natural process, *i.e.,* by evaporation and seepage. The amount of water that is given off from the site by these physical processes depends on a number of factors, such that in locations having strong sunshine, openness, high wind, low relative humidity, coarse or light textured soil, the amount of water which is drained out naturally is relatively high than when opposite conditions prevail. The water that is left after the natural drainage however, remains stagnated and evidently, this needs to be expelled out.

The common method of expulsion of this surplus water from the site is by making *drainage channels* which take up the water and pass it on to outside. Location of the drainage channels should however, be done judiciously so that all the water that is stagnated on the land whether at one part or more parts flows freely to the channels. The width and depth of the channels would depend on the existing condition and these should be regularly cleared of soil if deposited in them and weeds if grown there. In some areas of the southern and eastern Indian conditions where heavy rainfall occurs, trenches are dug instead of channels to drain out the surplus water in the rainy season, particularly in banana plantations. In places where water accumulates heavily, even pumping out of the water or removal with the help of *don* (syn. *duni, karin etc.*) becomes necessary. It should be remembered that water logged situation is detrimental to majority of the agricultural crops and perhaps more so for fruit crops.

The common symptoms that are noticeable in fruit crops under ill-drained conditions are yellowing of leaves, fall of flowers and fruits, collar-rot and anthracnose disease and prolonged stagnation even results in fall of the trees within a few days. Therefore, sooner the surplus water is drained out, better it is for the crop and also the soil, instead of procrastination in the matter. The decision that drainage should be done after a week or so to allow conservation of water in the orchard soil, or that it should be done when part of the stagnated water goes on by evaporation and seepage is never a prudential approach in fruit farming.

The selection of the irrigation method is no doubt, an important aspect as regard water management of the orchard is concerned. In heavy soil, flooding, check, ring and such free methods are not justifiable as these may hamper drainage by free percolation and hence, other methods as described earlier should be taken up with judgment. To irrigate in such type of soil, the general principle should be remembered, *i.e., to irrigate frequently but with less or very less amount of water.* By far, drip and sprinkler irrigation methods may be considered beneficial for soil of any texture and in undulated or slopy land. Whatever system or method that is followed, *regulated discharge of water* is most important and it should be judiciously decided keeping in view the soil of the orchard, crops grown there and the environmental conditions.

2. Quality of Water

Checking the quality of water before it is delivered to the orchard is an aspect that is useally overlooked but there is no denying the fact that this needs great attention, if the orchard is to be saved from the damaging effect of irrigation. Although the irrigation water that is used in greater part of India is safe or could be regarded to be almost satisfactory, harmful effect by the use of polluted water also very often comes to notice.

The pollutants in irrigation water may be silt or other inorganic or organic substances and these remain either in solution form or as undissolved, suspended particles. In general, soluble salts of sulphate, chloride and carbonate of calcium, magnesium, sodium, potassium *etc.*, are usually of more common occurrence in water which is lifted from underground source, while silt and suspended materials are found in water that comes from rivers, canals and tanks. The water which has been used to wash factories contains a number of pollutants, inorganic and sometimes organic and beyond doubt, this water causes serious injury to both the crops and the soil. Of late, presence of arsenic compounds sometimes becomes a problem with underground water in many parts of India.

Many fruit trees have the ability to tolerate impurities in water upto a certain limit. But when these become excess, the trees are badly suffered showing adverse symptoms, face moisture stress due to *exosmosis* phenomenon and may even be killed due to toxicity. The harmful effect is more serious in soil which may be permanently damaged due to change in the physico-chemical condition and also in destruction of beneficial biotic agents like earthworms and soil micro-organisms.

Among soluble salts, the carbonates are generally most harmful and the harmful effects are next with chlorides and then with sulphate salts. In general, it is not advisable to use water for irrigation if it contains 200 ppm (= parts per million) of chloride, 600 of sulphate and 100 ppm of carbonate salts, depending on crops to be irrigated. Presence of boron in the irrigation water is not desirable but toxic concentration of this element rarely occurs in the water used for irrigation under Indian conditions.

3. Water Contact at Tree Bases

In irrigating an orchard, care should be taken that the water does not remain in direct contact with the basal part of the fruit trees or shrubs for prolonged period. A fruit tree, papaya may be named which is extremely sensitive in this respect and the base of its trunk is quickly infected by the soil borne disease, *viz.*, collar rot, due to the fungus, *Phythium aphanidermatum*, if remains under water–logged condition. In citruses also, stagnation of water around the trunks and ill-drained situations are pre–disposing factors for spread of the fungal disease, gummosis due to *Phytophthora* species. The irrigation water may act as a carrier in the transmission of many other diseases to healthy plants. Particular mention may be made to the anthracnose diseases in grape which is caused by the *Gloeosporium* species of fungi and is very easily disseminated to healthy plants through irrigation water when it carries pruned out plant parts. The disease infections occur especially when the trunk of the trees or the branches are in direct contact with the contaminated water.

To protect the trunks against contact of irrigation water, many fruit growers pile up soil around the bases as shown in Figure 29. But this practice is never desirable as the basal part of the trees is liable to be infected more easily by the soil borne disease organisms, if remained covered with soil. In grafted trees, such soil mounds may come in contact with the graft operated part and thereby, may lead to infection at that part or may weaken the graft union.

Figure 29: Faulty Method of Protecting Tree Trunk against Water Contact by Heaping Soil Around the Base

Maintaining a slope from the base of the trunk and upto the border of the ring or the basin is done by fruit growers to protect the tree trunks (Figure 30). This method appears to be somewhat better as water logging does not take place near the trunk, besides that the water is also delivered beneath the fringe of the tree, *i.e.*, near to the water absorbing roots and gets accumulated there. However, this method has the drawback in that, in case of shallow rooted trees in particular, many of the feeder roots lying near the periphery of the basins are liable to be uncovered while the slopes are made.

A method which may be considered ideal is to make *two rings* instead of one around each tree. One ring should be made below the skirt of the tree or slightly away from it as is done in the ring or basin method of irrigation. The second ring should be of smaller diameter and this should be made close to the trunk, *i.e.*, at about 45 cm distance from the trunk, inside the basin or ring. Irrigation water should be applied in the space between the inner and the outer rings (Figure 31).

Figure 30: Protecting Tree Trunk Against Water Contact by Making Slope Towards the Periphery of the Basin.

Figure 31: Protecting Tree Trunk Against Water Contact by Making a Small Extra Ring Close to the Trunk.

By the above *double ring* method, moist soil as well as irrigation water does not come in direct contact with the trunk, for which it is saved from disease infection, roots are not uncovered and water is delivered close to the *feeder roots* of the tree, *i.e.*, underneath the peripheral spread of the tree.

4. Injury to Roots

While laying out the irrigation method, many roots and especially the feeder roots which remain close to the soil surface are sometimes injured or cut off. Due to that, water absorbing organs in the roots are reduced and disease infection takes place through the cut surface of the roots. The damage to roots results by deep working of the soil as in trench and furrow methods of irrigation. Such methods have therefore, little suitability for fruit trees and shrubs that are shallow–rooted, *e.g.*, papaya, guava, litchi, sapota, avocado, peach *etc.*

5. Irrigation During Pruning

Irrigation just before or just after pruning may cause harm rather than any benefit. It is in fact, not a scientific procedure to irrigate heavily just after pruning which is done by many fruit growers with the idea to make up the water loss that has taken place from the cut surface of the branches when pruned. This irrigation is unnecessary especially if done heavily. In grape, it is seen that if excess water is present in the soil after pruning, much of the water is flown outside the vine through the cut surfaces as if the water has been pumped out by the vine. This is known as *bleeding of grape vine* which is found in tobacco and some other plants also and is caused due to the *root pressure*, which is a physiological phenomenon.

6. Rootstock

There are some rootstocks in many fruit trees which have the ability to tolerate relatively high soil moisture while there are many which are just the opposite. Obviously, the trees growing over the latter rootstocks should be badly affected on receiving irrigation water in high amount and regulations of the amount of irrigation has a great necessity.

7. Irrigation in Dry Soil

Especially during a long spell of dry weather, sudden and heavy irrigation in the orchard when the soil is very dry is not only of any benefit but also becomes harmful to the trees. Sudden heavy uptake of water by the trees in their water stress condition may also result shedding of many leaves, flowers and fruits for abrupt water imbalance. Hence, heavy irrigation should not be done under such condition unless preceded by light irrigations.

8. Regulating Flow of Irrigation Water

Fast rate of movement of water through the irrigation channels may erode the soil and deposit it elsewhere, particularly in loose soil. This may also over-irrigate the area and result in water stagnation where drainage is poor. Therefore, the flow of water as well as the amount of water to be delivered should be judiciously regulated at the source in order to avoid such hazards and to make economy of irrigation water. This is of particular significance when irrigation water is desired to be delivered in the sub-soil.

9. Wastage of Water

Much amount of irrigation water sometimes goes into wastage in some orchards through leakages like rat holes, crevices, tunnels *etc*. To prevent such loss, release of irrigation water from the source alone is not enough but the orchardist should check up the course of irrigation water in the orchard and make suitable arrangements to seal the leakages wherever these are detected. Loss of water due to leakages in the channels may however, be saved by delivering irrigation water through pipelines, hose-pipes, polythene tubes *etc*. Lining the pipelines suitably with long polythene sheets may also prevent loss due to leakage. As irrigation is one of the costliest inputs in cultivation, unnecessary wastage of it is never justifiable.

10. Use of Drained Water

Where irrigation water is costly or not adequately available, many growers save the drained water, suitably store it and re–use the same for irrigation. The practice seeks favour as the nutrients which are washed out from the soil to the drainage channel may be recovered and supplied again. But this is not supported if the drained water carries leaves, twigs *etc*., which are disease infected.

11. Time of Irrigation

Irrigation during the midday, particularly in hot weather makes much wastage of water through evaporation. The trees also take up relatively high amount of water in higher soil temperature which is not desirable. Hence, irrigation in the morning or preferably in the afternoon hours seems to be the suitable time.

Part : II

ORCHARD SOIL MANAGEMENT PRACTICES

Chapter 12

Concepts of Orchard Soil Management

After planting of propagating materials in the orchard, adequate care is necessary to be paid both to the (i) growing fruit plants and the (ii) soil where they have been set. Effective measures on these aspects are necessary to be paid as long as the orchard exists in the location, in order that success in fruit farming is achieved.

Operations that are to be undertaken in the orchard soil are collectively known as *orchard soil management* or *orchard floor management practices.* A large number of activities are actually involved in the process and the following among them are of pertinence. (i) Conservation of optimum moisture, (ii) soil reaction (*p*H), (iii) micro–organisms, (iv) improvement of structure and its loosening for better aeration and percolation, (v) restoration of fertility, (vi) preservation of organic matter, (vii) control of weeds, (viii) control of soil erosion *etc.*

It needs mentioning in this context that much space is left between the fruit plants grown in an orchard. When they are in mature stage, much of the vacant space remains shaded by them partially or completely. But in early stages of their growth, area under shade is much less and during the first few years after establishment of the orchard, the shaded area is still lesser. This is common to all conventional orchards though may not be apparent in plantations where short duration fruit crops such as pineapple, strawberry, banana, melons, cape-gooseberry *etc.* are grown or high density or ultra high density method of planting is adopted. However, in these plantations also spaces are sometimes left when some particular methods of planting are adopted.

Now, if these spaces in the orchard are left as such without doing anything, there may be profuse growth of weeds. As the weeds take up considerable amount of moisture and nutrients from the soil, the fruit crops grown in the orchard would be badly suffered. The weeds also harbour many pests and diseases for which, the fruit crops may be attacked by those elements if there is abundant growth of weeds on the orchard soil.

Ploughing the orchard land facilitates manual removal of weeds besides that the soil is *aerated* for which the moisture is conserved in the soil and thus, the growing trees are benefited. But there is also the chance of injury of many roots by tilling the soil and thereby, this may affect health of the trees and the amount of water uptake by them.

Again, whether the vacant spaces left between the trees in the orchard be utilized for growing other crops leaves a question. This is because, by doing so, some extra production and thereby, extra earning from the orchard is possible. Besides this, growth of weeds is controlled in the orchard if other crops are grown there. On the other hand, these crops would withdraw much amount of water and nutrients from the soil and whether the fruit crops grown there would be suffered or not by them no doubt, needs attention.

Therefore, how these left out spaces in an orchard could be effectively managed, in other words, what judicious steps are to be taken on these spaces is an area that needs keen attention. This is an integral part of orcharding.

It should not be forgotten that the roots of the fruit crops have two vital functions, which are as follows:

(i) To take up moisture and nutrients from the soil medium and passing on the same to the above ground parts of the trees, and

(ii) To provide anchorage (= support) to the trees for their standing above ground.

Accordingly, in successful fruit culture, not only efficient management of the fruit crops but that of the land where their roots exist is also an aspect, the effective management of which can never be disregarded nor underestimated.

What management practice is suitable for the soil of a given orchard should in fact, be decided in consideration to a number of factors. It should be kept in mind that the system which is to be adopted in one orchard may not be applicable as such in another orchard. A number of principles however, becomes necessary to be taken into account for ideal management of the orchard soil and among these, the following are of greater pertinence:

(i) The first and foremost attention should be given to the fact that no such measure should be adopted which renders any type of harm to the growth of the fruit crops grown in the orchard and their productiveness.

(ii) Similarly, any measure which deteriorates physical, chemical and biological condition of the orchard soil that makes less conducive to the crops is not desirable by any means.

(iii) Such measures which help in bringing about higher productiveness of the fruit crops grown in the orchard should always get emphasis.

(iv) The measures that facilitiate orchard operations like tillage, manuring *etc*, have reasons to adopt.

(v) Securing an extra income from the orchard by raising other crops there needs encouragement, provided that this practice does not cause any harmful effect to the fruit crops as regard their productiveness.

Systems of Management

A number of orchard soil management methods are in fact, practised throughout the world. All these methods may be grouped under any of the following four systems:

(i) Without allowing any extra plant to grow in the orchard except the fruit plants for which it has been set up and without provision of any extraneous materials on the orchard land.

(ii) By growing other crop(s) or (non–crops) plants apart from the fruit plants of the orchard and without provision of any extraneous material on the orchard land.

(iii) Without growing any other plant except the fruit plants of the orchard but with provision of extraneous material(s) to over the orchard land.

(iv) Making a combination of one or more of the above systems either together or separately in blocks on the orchard land.

The methods that are commonly followed under the above systems and the merits and demerits of those have been discussed in the subsequent Chapters of the book.

Chapter 13

Management without Growing any Extra Plant (= Clean Culture)

In this system of orchard soil management, only those fruit crops for which the orchard has been set up remain present and without anything else. That is, no other plant whether crop or not is allowed to grow nor any extraneous material to cover the orchard land is provided. Accordingly, the spaces between the fruit trees are left vacant. As the land is virtually kept *clean, i.e.,* free from any plant except the fruit trees or any foreign material, the system is commonly termed as *clean culture*. The different methods that come under this system have been described in the following:

1. Clean Culture by Ploughing

The method is followed in most parts of India. In this method, the land is ploughed. Both deep and shallow ploughings are done at suitable intervals or at different seasons. Deep ploughing to about 15 cm is conventionally done during the winter season when the orchard trees are in minimal growth activity or have undergone dormancy as in case of deciduous fruit crops. This is done either by heavy type of plough, ridger *etc.* drawn by bullocks or disc harrow, cultivator *etc.* drawn by tractor. Ploughing with a soil-turning plough at least once or twice in a year is no doubt, of great advantage. Ploughing should be followed by cross ploughing and after that, the weeds are completely eradicated from the land. Levelling the land after ploughing is done by many orchardists although dispensed with by others. However, levelling is always desirable particularly for conservation of soil moisture.

Apart from winter ploughing, deep ploughing and cross ploughing the land just before and just after the rainy season is also recommended by many. Ploughing in this season is of advantage in conserving soil moisture and hence, it is of benefit in areas where rainfall is scanty or is not widely distributed over the year.

Apart from deep ploughing, shallow ploughing or even hoeing the land to a depth of less than 6 cm is needed to be done in this method several times in the year or at interval of 1 ½ to 2 months. The idea of this shallow tillage is to uproot any small growth of weeds that usually come up in between the heavy ploughings. Thus, the orchard is kept free from any weed at any time of the year as a result of the ploughings.

(a) Advantages

The following two advantages are especially derived by clean culture method:

(i) Weed Control

The weeds including grasses are controlled in most efficient manner in this method. Practically no weed or only a few weeds could be seen in an orchard where true clean culture has been adopted.

The harmful effect of weeds in an orchard needs no emphasis. These undesirable and unwanted plants withdraw large amount of moisture and nutrients from the soil and thereby, the fruit trees are suffered. Taking up large amount of available nitrogen from the soil by them is also of great concern as nitrogen is not very abundant in most Indian soils. The weeds also act as secondary hosts to many pathogens and harbour many insects and animal pests. The thorny and vigorously growing weeds interfere even movement of persons in the orchard.

(ii) Benefit of Tillage

The aeration of the orchard soil is increased greatly by ploughing. The microbial activity in the soil is thus, increased. By deep ploughing, the hardening of surface soil which may hamper impede percolation of water is avoided and besides that, deep ploughing ensures percolation of water to a greater depth of soil.

(b) Disadvantages

The clean culture by ploughing has however, the following demerits:

(i) Exposure and Damage of Roots

Many roots of the orchard trees lay bare or damaged or are cut off when ploughing is done. As a result, root rot disease is favoured, uptake of water and nutrients from the soil may be reduced besides that there may be poor anchorage for the fruit trees or the shrubs. Due to damage or destruction of roots of the fruit plants, the method is hardly recommended now, except in some condition of the soil. Destruction of feeder roots which are the absorbing organs of the fruit plants causes lesser uptake of water and nutrients from the soil for which their growth is stunted.

(ii) Depletion of Soil Nitrogen

Due to high aeration of the soil, nitrogen level in it is depleted to a considerable extent.

(iii) Formation of Hard Pan

The method leads to formation of hard pan in the soil.

(iv) Reduction of Humus

Frequent cultivation depletes humus matter from the soil.

(v) Frost Injury

As the soil lay bare by cultivation, roots of the fruit plants are more prone to frost injury in winter season.

(vi) Cracking of Soil

Especially where the soil is of clayee nature, it undergoes cracking easily in summer season by adopting this method.

(vii) Soil Erosion

Frequent cultivation brings about more erosion of soil in uneven land.

(viii) No Extra Production or Income

No supplementary crop is grown in the clean culture method of soil management. Hence, no extra production or income is possible from the orchard.

2. Clean Culture by Application of Weedicides

In this method, the land is not ploughed and the weeds are controlled by application of *chemical weedicides*. Nevertheless, hoeing the orchard land lightly at suitable intervals is also sometimes practised to uproot the stubborn grasses and such weeds which are difficult to be controlled by weedicides. In under to control weeds, some orchardists even burn the weeds by pouring kerosene over them. But this a very harmful practice. (N. B.: When something is burnt on an agricultural land, many micro-organisms and earthworms present in the soil are also killed. Hence, productivity of the crop plants is greatly reduced as these organisms enormously help in crop production. Even decades may be required for re-emergence of micro-organisms. Hence, burning anything on agricultural land is prohibited by law by the Govt. India).

(a) Advantages

The method has its advantage in that the damage of roots as may result by tillage is avoided and labour cost due to that is saved. The method is widely practiced in many countries in the world in large sized orchards, if inter cropping is not desired.

(b) Disadvantages

The surface soil may get hardened in this method if hoeing the land is not done at regular interval. Any extra production or income by growing supplementary crop in the orchard is also not possible.

Chadha in his book, *Handbook of Horticulture* has presented a list of chemical weedicides having effectiveness in different fruit crops and the table is presented below.

Fruit Crop	*Recommended Herbicide*	*Dose (Kg/hectare)*	*No. of Spray*	*Time of Spray*
Mango (Nursery)	Diuron or	2.25	1	Just after sowing seeds
	Oxyfluorfen	4.00	1	Just after sowing seeds
Mango orchard (only in first 5 years)	Diuron or	2.25	1	Just after transplanting
	Atrazine or	4.00	1	Just after transplanting
	Oxyfluorfen	4.00	1	Just after transplanting
Banana	Diuron or	2.25	1	Just after planting suckers
	Simazine or	4.00	1	Just after planting suckers
	Oxyfluorfen or	4.00	1	Just after planting suckers
	Paraquat or	12.00	2	6 months after pre–emergence spray
	Glyphosate	6.00	1	6 months after pre–emergence spray
Pineapple	Bromacil + Diuron	2.25	1	10–15 days after planting
Grape (Nursery)	Oxyfluorfen or	4.00	1	Just after planting
	Diuron	2.25	1	Just after planting
Grape (Vineyard)	Atrazine or	4.00	1	Just after pruning during April
	Diuron	2.25		
	Paraquat or	12.00	2	As and when weeds emerge and become 15–25 cm tall
	Glyphosate	6.00		
Papaya (Nursery and orchard)	Fluchloralin or	4.00	1	1 week after transplanting
	Butachlor or	4.00	1	1 week after transplanting
	Alachlor	4.00	1	
Papaya(orchard only)	Glyphosate	6.00	2	5–6 months after pre–emergent spray
Citrus (Nursery and Orchard)	Diuron	5.00	1	After sowing seeds and transplanting
Guava (Orchard)	2, 4–D	2.00	1	Established orchards Twice yearly during rainy and winter season
	Paraquat	6.00	2	
Guava (Nursery)	Diuron	0.60	1	After sowing
Litchi	Paraquat	6.00	2	One before and one after 2 months of rainy season
Apple	Atrazine or Simazine or Diuron	5.00	1	After harvesting and earthing up

Fruit Crop	*Recommended Herbicide*	*Dose (Kg/hectare)*	*No. of Spray*	*Time of Spray*
	Paraquat	3.00	2	5–6 months after pre–emergent spraying
Peach	2, 4–D or	2.00	1	After harvesting fruits
	Simazine	5.00	1	After harvesting fruits
Apricot	Diuron or	2.00	1	After harvesting fruits
	Simazine	5.00	1	After harvesting fruits
Plum	Diuron or	5.00	1	After harvesting fruits
	Simazine or	4.00	1	After harvesting fruits
	Atrazine	4.00	1	After harvesting fruits
Strawberry	Simazine or	2.00	1	After planting
	Chloroxuron	2.00	1	After planting
Cherry	Simazine	1.00	1	After harvesting fruits

3. Clean Culture by Mud Plastering

This is a method which is largely found in the central and southern parts of West Bengal, especially in the mango orchards. The orchard land is made even as much as possible by soil work and levelling. Then the land is plastered by applying fresh cow dung and clay. Plastering is done so carefully that when dried, the land appears to be a floor of the cottage yard. The plastering is repeated at suitable intervals except in rainy season.

Such cow dung plastering popularly known as *nikano* method is in fact, done for some specific objectives. These are, when the orchard land is used for threshing grains, for shelter of domestic animals, for storing hay and dry fodder and also as a place for community meeting, shelter of people *etc*. The small orchards are usually used for this purpose and not the large sized ones.

(a) Advantages

By ploughing the soil, a mulching effect is produced on the orchard land. The soil moisture is thereby, greatly conserved as no weed is grown and the soil is covered with a layer of cow dung. Besides, this, the shady place of the orchard is utilized for other purposes as well.

(b) Disadvantages

Manuring and irrigation are difficult in such orchards as the mud–plastering is done almost throughout the year except in rainy season.For this difficulty, many growers adopt plastering for some seasons only. No supplementary income is also possible from the orchard land by this method by growing other crop(s).

Chapter 14

Management by Growing Economically Important Plants (= Crop Culture)

This is a system of orchard soil management which consists of growing other crop(s) in the orchard, in addition to the fruit crop for which the orchard has been set up. Only other crop(s) is grown in this system but no material to cover the orchard land is provided.

The other crops that are grown in this system include food crops that are consumable to the human beings or other economically important crops which include forage crops also. In accordance with, the fruit crop for which the orchard is set up is termed as the *main crop* or the *principal crop* or the *principal fruit crop* and the other crops as *supplementary crops, secondary crops, subsidiary crops, supporting crops, side crops, extra crops etc.*

These supplementary crops could be divided into two groups, which are:

(i) Secondary fruit crops, and

(ii) Other economically important crops.

These have been brought forward in the following:

1. Growing Secondary Fruit Crops (= Fillers)

In this method, in addition to the principal fruit crops, other *fruit crops* are grown in the orchard. The latter fruit crops are therefore, regarded as *secondary* or *supplementary fruit crops* to distinguish those from the *principal, i.e.,* the *main fruit crop*. However, these secondary fruit crops are most commonly termed as *fillers.*

The term, filler is perhaps inappropriate, because if it denotes that the fruit crop which fills the vacant space of the orchard should be called fillers, then many other crops or non–crop plants or other extraneous materials (stated later) which also do the same, *i.e.*, make filling should be termed so. But these are not called as fillers. Nevertheless, the term, filler is of more common usage in horticultural literature and hence, this term is mostly retained in this text.

It needs mentioning in this connection that growing fillers in the orchard is usually considered as a method of fruit culture and accordingly, the subject is brought forward under the systems of orchard layout rather than as a method of orchard soil management. But as fillers is not a must in an orchard and as it is grown in the inter–spaces in an orchard, it is perhaps more relevant to consider it as a soil management practice.

However, if it is desired to grow fillers in the orchard, the following considerations should be kept in view.

(i) Only some particular species or even varieties of fruit crops are recognized as fillers and these are selected for growing taking, into consideration a set of conditions. These conditions are, they may be tree, shrub or herb, dwarf in height such that their height do not exceed the main (= principal) fruit crop, compact in growth with less spread of sidewise growth, their roots do not extend to long distance, not voracious feeder, *i.e.*, do not take up heavy amount of nutrients and water from the soil and above all, do not cause any harm to the main fruit crop as regard, health, growth and productiveness of it both in yield and quality of the produce. The fillers are termed as filler trees, shrub or herb, respectively as they are trees, shrub or herb.

The fruit crops that are conventionally grown as fillers include those like papaya, banana, *phalsa,* pomegranate, *ber*, peach *etc*. Apart from these, other fruit crops but of dwarf varieties are also grown, *e.g.*, dwarf varieties of guava, jackfruit, lime *etc.* but to a lesser extent. Some dwarf and compact minor fruit corps like *karonda*, mulberry, calamondin (= *hazara)*, Surinam cherry, West Indian or Barbados cherry and in temperature zone, some wild types of berries and some other fruits which are not cultivated in organized orchards are also sometimes selected for growing as fillers. Even grape vine trained on single stake system is seantily found to grow as a filler in some mango orchards in the Punjab. In any case, the crops grown as fillers should produce such fruits that have edible value to the human beings either as raw or after making products.

The question arises in this connection, that is, whether pineapple, strawberry, cape–gooseberry, water melon, musk melon *etc.* which do not reach a height of more than 40 cm or are creepers should be regarded as *fillers* or not. Many horticultural authorities have considered these as fillers because they come under fruit crops. Mazumdar in his book, *Principels and Methods of Orchard Establishment. has however, remarked* that as these crops are of very short height and as these are cultivated like annual crops, it would be more appropriate to consider these crops as intercrops rather than as fillers.

(ii) More commonly, the filler trees, shrubs and herbs as the case may be are cropped in the orchard *temporarily, i.e.,* for a short duration only. Hence, these are also termed as *temporary fruit crops* in the orchard. However, it should be noted that they are always not grown temporarily and are sometimes grown permanently also.

(iii) Filler trees, shrub or herb may be grown at any stage in the life of the orchard, *i.e.,* from the time of its establishment and till it lasts.

(iv) The general convention is to set the filler plants in the inter–spaces of the main fruit trees in the system of orchard lay out which is known as *square system* of lay out. That is, lay out of the orchard is done following square system [Figure 32(a)] by making with lime. Pegs are infixed on the ground at each corner of the squares drawn. These pegs indicate the locations where the propagating materials for the main, *i.e.,* permanent fruit crops are to be planted. Then the middlemost position of each individual square is determined, which is done on finding out the point of intersection of the two diagonal lines drawn by joining the opposite corners of each square [Figure 32(b)]. The intersecting point of the two diagonal lines in each square indicating the middlemost position is pegged [Figure 32(c)] and these pegs in the individual squares indicate where fillers are to be planted. This system of layout is in fact, known as the *diagonal* or *quincunx system of layout.*

However, it may be noted that although fillers are conventionally set in the diagonal system of orchard layout, they may be grown in any other systems of lay out as well.

Method of Cropping

An important point should never escape attention of the orchardist, that is, the *orchard is meant for the principal (= main) fruit crop* and has not been established for the fillers or the inter–crops.

Thus, any crop grown in the orchard other than the main fruit crop should be looked upon as a supplementary or subsidiary crop only and that, it is not a must in the orchard unlike the main fruit crop, existence of which is essential. In accordance with, *attention to the main fruit crop should be much greater* to see that they do not suffer by growing fillers. To ensure this, the following steps should particularly be adopted in cropping of the fillers.

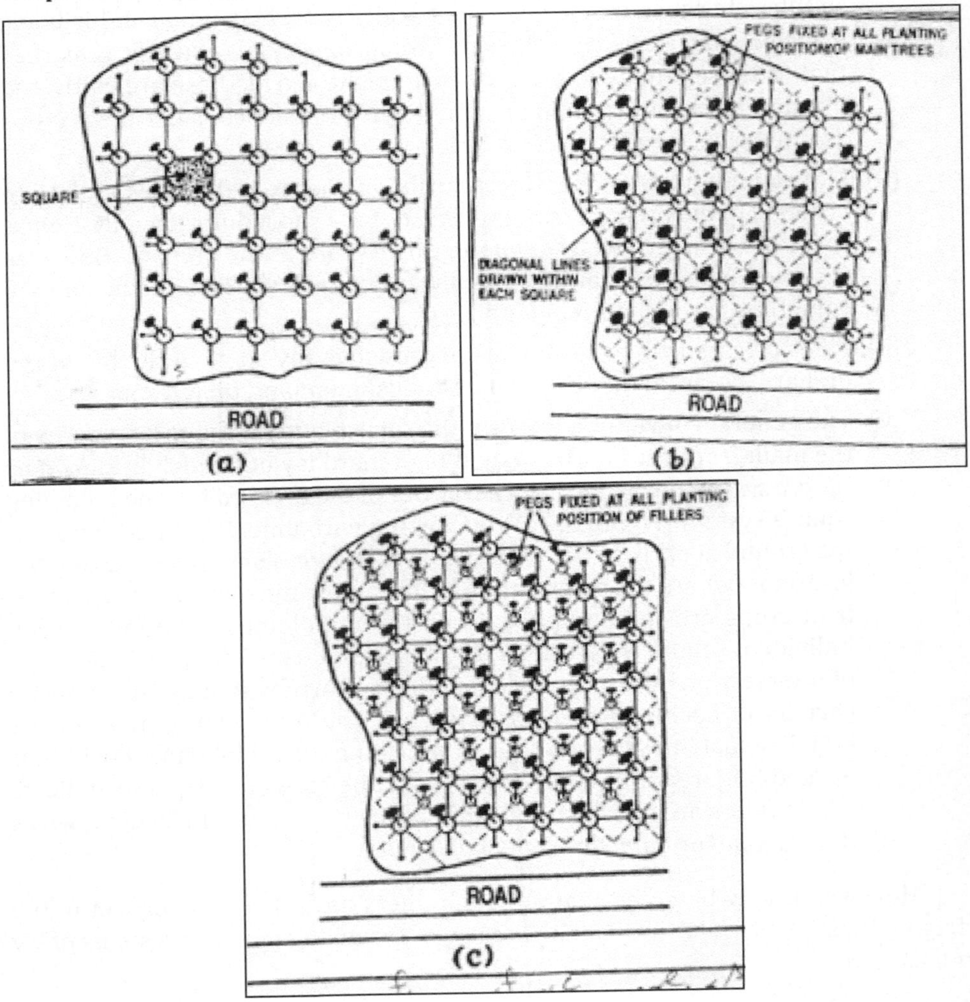

Figures 32: (a) Pegs fixed at four corners of each square in square system of layout (b) Diagonal lines drawn within each square (c) Pegs fixed at the point of intersection of diagonal lines to indicate position of fillers.

(Courtesy. *Principles and Methods of Orchard Establishment* by Dr. B.C. Mazumdar, 2003, Daya Publishing House, New Delhi).

The first and foremost step is to make arrangement to prevent *competition* between the main fruit trees and the fillers as far as possible, both in their underground and the above ground parts. The underground competition takes place between the *roots* of the fruit trees and the fillers as regard uptake of moisture and nutrients from the soil as well as for space. For such competition, not only the main fruit trees but the fillers may also be deprived in taking up the required amount of moisture and nutrients from the soil. Besides, this, there may be a crowding of roots resulting poor growth of them.

To avert such type of undesirable competition, manure, fertilizers and irrigation are to be applied *separately* for the main fruit crop and the fillers. That is, the amount of irrigation and the kind and amount of manures and fertilizers that are to be applied should be decided taking into account the actual requirements for both these two types of crops. Besides, the main application of these inputs should also depend on the actual time of their requirement. Such judicious application of the inputs is expected to keep off competition in the underground sector. This aspect needs keen attention particularly as nothing is visible what takes place in the underground.

The above ground competition implies struggling between the branches of the main fruit trees and those of the fillers in respect of space. It is sometimes noticed that the fillers make prolific vegetative growth and send forth the branches to such an extent that overlapping of branches of the main fruit trees and the fillers result. Such crowding of branches hamper growth and productiveness of the main fruit trees and also to the fillers. To avoid this competition, it is necessary that such varieties of filler crops should be selected for growing which are relatively *dwarf* and *compact* in growth. Besides this, it is also necessary that they should be properly *trained* and *pruned* so as to check excessive vegetative growth of them. For example, if fillers comprise as trees, they should better be trained to modified central leader or the central leader system rather than the vase (= goblet) system so that the sidewise canopy spread of them is restricted.

If banana is grown as a filler, regular *pruning of the suckers* of this herb is a very important step. The operation which is known as *desuckering* should preferably be done by the method in which no extra sucker is allowed to grow till the main plant starts flowering, *i.e.*, any sucker emerging before that should be cut off from the rhizome with a cutlass or a crow-bar having a chisel like end. As flowering progresses, one sucker should be allowed to grow and the rest are cut off. Any other sucker coming after that should be cut off till the mother plant is in fruiting stage when another new sucker is allowed to grow and the rest are cut off. When the mother plant dies after harvest of the fruit bunch, the first sucker which will be about five months old at that time takes the lead and becomes the next mother plant. The new suckers that emerge should be cut off till the first sucker starts flowering when another new sucker is allowed to grow and the rest are cut off.

In this way, the operation is repeated such that only one new sucker is allowed to grow every year and the rest are cut back. Accordingly, not more than two suckers apart from the mother plant (= fruiting plant) are present in a clump throughout the year. If left to nature without desuckering, numerous sucker may be produced from the rhizome and these will take up large amount of moisture and nutrients from the soil. As a result, growth and fruiting performance of the mother plant will be hampered and the main fruit trees in the orchard will also face competition.

The condition is more serious in those varieties of banana, the suckering tendency of which is very high.

Another essential step for the fillers is to take adequate plant protection measures of them. It is a fallacy that the control of pests and diseases should be done only for the main fruit trees in the orchard and not for the fillers. This is because, the pests and disease organisms which infect the fillers may infect the main fruit trees also if not timely controlled and the fillers may be secondary hosts to many insect pests and pathogens.

(a) Advantages

The following advantages are no doubt, derived by growing fillers in the orchard:

(i) Number of fruit crops is greatly increased in the orchard. The number may be double when set in diagonal system of orchard layout.

(ii) More than one species of fruit crop is grown by planting fillers. Accordingly, return from the orchard is also higher.

(iii) As more than one species of fruit crop is grown in the orchard, more than one species of fruit is available. Due to that, if there is any risk or loss from the main fruit crop in any year, that is compensated by the fillers.

(iv) Weed growth is much less in the orchard as most part of it remains shaded, either by the principal fruit crop or the fillers.

(v) Utilization of the vacant spaces in the orchard is effectively done by growing filler.

(vi) An interesting feature of growing fillers in an orchard lies in the fact that there occurs a mutual dependence between the principal fruit trees and the fillers. That is, roots of the fillers usually do not forage from that part of soil where the roots of the principal trees withdraw large amount of moisture and nutrients. Again, at seasons when the uptake of nutrients is high by the one crop, the other crop takes up relatively less amount from the soil. The above-ground foliage growth of one is also adjusted in conformity with growth of the other. But such coordination takes place when selection and culture of both the principal fruit trees and the fillers are done in an effective and intelligent manner.

(vii) Filler trees are sometimes recognized as *risk avertible trees* in the orchard. That is, if income from the orchard is insufficient, the filler trees compensate the loss. For example, mango trees which are threatened by the problem of alternate or irregular bearing habit give nil or scanty fruits when they are in "off" years. But production from the mango orchard is possible if filler trees grow in it.

(b) Disadvantages

Despite the above advantages, there may be some drawbacks in growing fillers in an orchard which may be stated as follow.

(i) When the main fruit trees and the fillers are in grown up condition, little space is left in many cases between the crowns of the former and the latter trees. As a result of the crowding, ventilation, *i.e., drainage of air* in the orchard is greatly hampered. This is particularly apparent in temperate climate and in winter season in many tropical and sub–tropical situations. For this poor air drainage and trapping of cold air, growth and yield performance of both the crops are reduced. For this reason, many horticultural scientists do not recommended growing of fillers in the orchard, except in places where such situation is not likely to arise.

(ii) Due to crowding of trees and thereby, less availability of space, harvesting, spraying and other operations to the trees may have difficulty.

(iii) If adequate irrigation, manure, type and amount of fertilizers, crop protection measures are not applied to the fillers and in time when these are needed, the main fruit trees may be seriously suffered. Similarly, excessive vegetative growth if any of the fillers is not checked, the growth and fruiting performance of the main fruit trees may be hampered.

2. Growing many Species of Fruit Trees in same Orchard

In many parts of India, there is a trend to grow a number of species of fruit crops in the same land. In this system, all fruit trees in the orchard are principal, *i.e.*, main fruit trees and no tree is recognized as filler.

There is advantage in this type of planting as fruits belonging to many species could be obtained from the same orchard. However, there is disadvantage in that, prolific growth of one tree may suppress the growth of the neighbouring tree if the latter is a less vigorous one. Orchard soil management may also have difficulty in such an orchard.

3. Growing other Secondary Crops of Economic Importance (= Intercrops)

In this method, the orchard soil is managed by growing *economically important crops* except fruit crops. Thus, the crops that may be grown include vegetable crops, spices, condiments, pulses, oilseeds, forage crops, fibre crops, floricultural crops, medicinal and aromatic crops, sugarcane, sugarbeet, mulberry, narcotic crops *etc*. The cereals and millets may also be cropped where possible but these are grown to a lesser extent. Fruit crops are not considered in this method but as stated above under the heading (1), some fruit crops, *viz*, pineapple, strawberry, melon *etc*. which are of much low height and are grown like annual crops should according to Mazumdar (2003) be included in this method.

Method of Cropping

The basic principle in growing intercrops in the orchard should be kept in mind as has been mentioned earlier in cropping with fillers. That is, the fillers and the intercrops are only supplementary members in the orchard unlike the main fruit trees of the orchard which is the principal crops. Accordingly, adequate care should be taken in their culture to ensure that the main fruit crop is not suffered in any way. The following measures are particularly mentionable when the orchardist desires intercropping in his orchard.

(i) Location of Growing

Broadly speaking, the intercrops are grown in the vacant space in the orchard in between the main fruit trees. However, to allocate these spaces, the orchardist should make the planning even much earlier, *i.e.*, before setting up the orchard. In other words, if he has the intention to grow intercrops as long as the orchard exists, he should layout the orchard following some particular systems. For example, rectangular system of layout provides *more vacant space* between the rows unlike the square, triangular or the hexagonal systems and hence, for long term intercropping, the former system appears to be more suitable. If layout by the latter systems is done, it may be necessary to increase the planting distance so as to provide somewhat more space for the intercrops to grow.

The above considerations become necessary for two reasons. These are, firstly, if minimum amount of land, especially unshaded area is not available, most of the intecrops will have poor performance which does not justify their growing at all only to make some unnecessary expenditure and secondly, unless allowance is given between the main fruit crop and the intercrops, both will suffer. Therefore, it is a justifiable approach to plan beforehand if intercropping is to be practised permanently.

However, the above viewpoint does not necessarily mean that intercrops will give poor production of them or will produce adverse effect to the main fruit trees if the latter are planted by square, triangular or hexagonal systems of layout and providing normal planting distance. In such systems of layout, the intercrops should in fact, be grown in such a way that their roots remain sufficiently away from the roots of the fruit trees. This may be accomplished on making the following considerations. It is stated that the roots of a tree extend upto that area under the ground to which the above ground parts have actually extended. That is, as shown in Figure 33, the roots will extend upto that part under the ground where the vertical (= perpendicular) lines if drawn from the skirt, *i.e.*, the peripheral spread of the tree touch the ground on all sides.

Accordingly, when the tree is very young, *i.e.*, after its planting, very less area is occupied by the roots and as the above-ground parts gradually increase, the zone occupied by the roots goes on increasing. In this way, as any large increase in the above-ground (= foliage) part does not occur when the tree reaches optimum maturity, no further extension growth of the roots also does not take place. Thus, the soil zone covered by roots will be lesser when the tree is in young stage and this will progressively increase every year to reach the maximum when the tree would attain optimum maturity condition.

Figure 33: Growth of Roots of a Tree in Conformity with Peripheral Spread of the Above Ground Parts.

However, it needs pointing out in this context that although the above relationship between the above-ground and below-ground parts are found to be true for some shrubs and trees but this is not for all. In fact, for mango and many other trees, it is observable that the roots have extended to long distance crossing the peripheral zone of the trees. Nevertheless, if the former viewpoint is considered, it should be opined that no intercrop should be grown within the peripheral zone of the main fruit tree in the orchard. Although, theoretically this may be acceptable, in practical sense, it is advisable that not only within the peripheral zone of the tree but even within some distance away from that zone, no intercrop should be grow. Leaving this zone, intercrops could be freely grown in other open spaces. But again, it is pointed out that provision for paths, irrigation and drainage channels have also their necessity in the orchard. This aspect should no doubt, be kept in view in allocating space for the intercrops.

The above discussion makes it imperative that the roots of the inter–crops should not be present in places where the roots of the fruit trees forage. In other words, any competition between the roots of these two types of crops needs to be avoided. The danger of competition in the above-ground parts usually does not take place between the intercrops and the fruit trees as may happen in case of fillers for the reason that most of the common intercrops that are grown are much dwarf in height and remain at much lower level than the crown of the fruit trees.

(ii) Selection of Species and Varieties

While arbitrary selection of the crops without any basis makes more harm or produce no benefit, judicious selection brings about success in intercropping. It should be remembered that not only wise selection of the species of crop plants but even variety is also an important consideration. In this selection of intercrops, a number of considerations should in fact, be taken into account, such as adaptability to the existing soil and climate, suitability to grow and produce well with the amount of available light and its duration during the day and in seasons, whether produces any harmful effect or not to the fruit trees, requirement that exists in the market, profitability *etc.* and the success depends on these factors.

In shaded situation or where availability of light is less, crops like turmeric, ginger and other shade–loving plants and even betel vines may be considered for growing . In sub–tropical parts of northern India where summer temperature goes high and winter is associated with frost, the tropical fruit crops, pineapple may be tried, especially the *Kew* types of pineapple under the shade of the trees. In shady situation, the size of the fruits of this type is known to be higher although sugar formation is lesser. Lemon grass, asparagus, gladiolus and some medicinal plants may also be grown in partial shade.

The selection of pulses to grow as intercrop gains much support as their roots add nitrogen to the soil apart from yielding grains or fodder. If pulses are not grown separately in plots, they may even be interplanted with other crops.

Some growers use the branches of the fruit trees as the support for growing the creeping vegetables like the cucurbits. This is however, not a good practice as the foliage parts of the fruit trees become shaded by those of these vegetables and hence, growth of the former may be hampered. It is advisable to provide stakes or scaffolding arrangement for the purpose of creeping of these crops.

The *rotation of crops* is an aspect that needs adequate consideration in raising intercrops in the orchard. It may be noted that properly followed rotation makes much improvement of the soil beside providing variety of crops in different seasons.

Many of the vegetables and other crops are known to be susceptible to viruses and other diseases. As fruit trees are liable to be infected from these crops if the latter are infected with diseases, it is advisable not to grow these crops or grow only disease resistant varieties of such crops. Some of the phanerogamic parasites are also known to be transmitted from some vegetables as *Orobanche* in tomato, brinjal *etc.* when these are grown as intercrops and hence, the orchard should guard against those.

(iii) Application of Manure, Fertilizers and Irrigation

As stated in fillers, these agro–inputs should be applied separately for the intercrops to avert any competition with the roots of the fruit trees. In other words, the irrigation and manure and the kind of fertilizers must be applied separately to the plots where intercrops are grown keeping in view the time of their requirement and the amount.

(iv) Crop Protection Measures

Necessary crop protection measures must be taken for the intercrops as in absence of that, they may be attacked by pests and diseases and from them, the fruit trees in the orchard may also be affected (= opportunistic infection).

(a) Advantages

The most important advantage of raising intercrops in the orchard is to secure an extra income from the orchard land. Especially during the early life of the orchard when the fruit plants are in primary stage of growth, enough space is available and extensive intercropping could be successfully done without causing any harm to the fruit plants. The data available from many parts of India have provided evidence that the cost of establishment of an orchard and the recurring expenditure during the first few years of it could be entirely met up if intercropping is done in an intelligent manner during this period.

Besides the above, textural condition and fertility of the orchard land are also improved by intercropping, especially when done during the first years. Intercropping with pulses adds adequate nitrogen to the soil. Weed growth is controlled by intercropping besides that the fallen leaves of these crops act as a partial mulch cover to the land. Any possible risk or loss of the main fruit crop in any year is also compensated by the intercrops.

(b) Disadvantages

If not judiciously done, intercropping may cause serious harm to the fruit trees and also the orchard land. The horticultural experts therefore, advise not to practise intercropping in the orchard if it is not possible to practice in an effective and intelligent manner. They should remind the orchardist, "Grasp at the shadow and lose the substance".

Chapter 15

Management by Growing Soil Conserving Plants (= Cover Cropping)

Agricultural soils comprise thin layer of Earth's crust where crop plants grow. They are formed after thousands of years of physical, chemical and biological weathering of rocks and hence, are highly precious material. To add much importance to soil, the United Nations had declared December 5 as "World Soil Day". Unfortunately, much of the soil on the Earth are lost every year. Although they remain on the Earth and are not vanished but change their position and thus, remain unavailable to the crop plants. Loss of the soil takes place due to the phenomena of *erosion*. This is a natural process which involves *wearing away of the surface soil by the action of water and wind.* Erosion by water takes place due to rain, stream, river, flood, seasonal flow of water from one place to another *etc.* Washing away of the surface soil takes place largely in hilly areas and in uneven topographic condition. The erosion of soil in the form of sheet takes a more serious turn in course of time to form rills and gullies resulting much greater loss of soil. Soil erosion by wind results in blowing off of the surface soil in the form of particles due to high wind and is more common in arid and semi–arid region in India. Wind brings about erosion as saltation, suspension and surface creep. Water erosion is caused by precipitation, slope of land, type of soil, ground cover and use of land.

Prevention of Soil Erosion in Orchards

Franklin D. Roosevelt remarked, "A nation that destroys its soil destroys itself". Prevention or retardation of soil erosion could be done by mechanical means or effective agricultural practices. Agricultural procedure involves deep tillage, conservation tillage, contour farming, mixed cropping, strip cropping, growing cover crops *etc.*

Cover crops are those crops which bind the soil and protect it from erosional loss. Evidently, such plants which are compact and bushy so as to form a protective vegetation cover on the soil surface and the roots are highly branched and they are able to bind the soil effectively. They are desirable materials to conserve the soil against erosional hazard. Therefore, such type of plants should be preferred for growing as cover crops to get the desired result.

Plants Suitable to Grow as Cover Crops

A number of crop or non–crop plants offer as suitable materials for growing as cover crops in the orchards. These include legumes used for green manuring or grains such as *dhaincha (Sesbania aculeata),* sunnhemp (*Crotolaria juncea*), lentil *(Lens esculenta)*, cluster bean (*Cyamopsis tetragonoloba*), Bengal gram (*Cicer arietinum*), green gram (*Phaseolus aureus*), cowpea (Vigna species), groundnut (*Arachis hypogea*), *etc.*, fodder crops like *berseem* (*Trifolium alex andrianum*), lucerne (*Medicago sativa*), napier grass (*Pennisetum purpuram*), Guinea grass (*Panicum maximum*) *etc.*, some cereals and millets like wheat (*Triticum locativum*), bajra (*Pennisetum typhoides*), *jowar* (*Sorghum vulgare*), maize (*Zea mays*) *etc.*, some floricultural plants like marigold (*Tagetes* species), ixora (*Ixora* species), chrysanthemum (*Chrysanthemum* species) *etc.*, some grasses like Bermuda grass (*Cynodon dactylon*), mutha (*Cyperus rotundus*), *Phalaris* species, *Chrysopogan montanus etc.* and many wild plants may also be successfully grown for the purpose. Which plant or plants are to be grown would evidently depend on the soil condition, climate, type of erosion and many other factors.

Method of Growing

Cover cropping may be done on the entire soil surface of the orchard. More commonly, instead of growing over the entire surface, these are grown in *strips, i.e.,* grown between two rows of fruit trees and the adjacent inter-row space is left as such without growing the cover crop. This pattern is maintained over the entire orchard area. In this type of arrangement, many orchardists prefer to grow cover crops in the inter-row alleys in alternate years. That is, in one year the cover crop is grown in an inter-row space while during the next year, it is omitted from that place and is grown in the adjacent strip which was left as such in the previous year.

Whether the cover crop should be grown throughout the year or in some particular seasons only should be decided in consideration to the extent of erosion and the cause of it. In some parts in India, it becomes necessary to grow these crops in most parts of the year but in other areas, these are usually grown only during the rainy season when danger of soil erosion due to water is greater. When grown in rainy season, the roots of the fruit trees are also less subjected to competition with them in respect of soil moisture. In areas threatened by blowing of surface soil in particles due to wind in winter, the cover crops may be grown in this season also. In the summer season however, cover cropping is usually not practised as they compete with the fruit trees for soil moisture, besides that the necessity is not felt much.

How the cover crop is to be grown in the orchard is a matter that has in fact, difference of opinion. One group believes that these should be grown just by broadcasting the seed without resorting to any tillage or application of agro inputs. This is however, considered only to be a fallacy by the other group, who does not support this type of culture. That is, according to them, a minimum agronomic practice is essential for their culture without which there will be a failure of the crops. Undoubtedly, the second group deserves support and one or two ploughings with application of manure and desired type of fertilizers are indispensable to raise the crop successfully. Occasionally, spraying the crop separately with aqueous solutions of low concentrations of urea, salts of zinc, magnesium, and potassium *etc.* gives encouraging results to boost up their foliage and root growth. Needless to say, crop protection measures to control pests and diseases are essential not only to save the cover crop themselves but also the fruit crop in the orchard.

(a) Advantages

The greatest advantages of growing cover crops lies in the fact that it conserves the orchard soil against erosion. This seeks justification for growing it in those places though not everywhere. However, apart from this, they offer many other advantages also. For example, when leguminous plants are grown, they add nitrogen to the soil and pulse grains to the human beings. When forage crops are grown, they provide animal feed. They protect the soil against high or low atmosphere temperature as well. Many plants belonging to this group are even able to retain sufficient amount of water in their leaf-blades and axils after rainfall and this is passed on to the soil. Sufficient amount bio–mass is also known to be added in the soil by most of the cover crops. In the U.S.A., it has been observed that by growing some cover crops in the apple orchards, "a significantly lower incidence of insect pests has resulted, mainly because of increased abundance and efficiency of predators and parasitoids supported by alternative prey or hosts harboured by the cover crops".

(b) Disadvantages

The greatest disadvantage of the cover crops is, they take up heavy amount of moisture and nutrients from the soil. Therefore, in soils of low moisture and fertility, the fruit crops grown in the orchards is found to suffer badly due to deprivation of moisture and nutrients. Some cover crops harbour many insect pests and diseases as well. However, the necessity of growing cover crop should not be disregarded for these drawbacks in areas where it has essentiality, particularly as the constraints are possible to overcome.

Chapter 16

Management by Growing Grasses (= Sod Culture)

Sod means a turf, *i.e.*, a lump of earth on which grasses grow. In *sod culture* (= turf culture) method of orchard soil management, grasses are allowed to grow all over the orchard land however, leaving the space just below the canopy of the fruit trees, which is kept clean and free from growth of any grass or vegetation. The method is more commonly practised in cool and moist regions of other countries than under Indian condition.

The grasses that grow naturally on the land may be maintained or grass seeds may be broadcast. The latter method appears to be superior to bring about uniform coverage and more compactness of sodded area. In an ideal sod, the grasses should be shallow rooted, low heighted and of bushy growth. They should also have low requirement of water. A number of species may be used but there is a preference of Bermuda grass (syn. Bahama grass) which is also known as *Doorba* or *Doob* (*Cynoden dactylon*) in most parts of India where sod culture is practised in the orchard.

Normally any tillage, manuring, irrigation *etc.* is not done in growing grasses but eradication of weeds or undesirable grasses become necessary. Some orchardists occasionally sprinkle highly diluted cow dung to boost up the growth of the grasses. Sprinkling low concentrations of solutions of urea and magnesium salts separately at suitable intervals also brings about abundant vegetative growth of the grass in short time. Sometimes patches of non–grassy area are visible here and there due to disease infection which need to be controlled by suitable plant protection chemicals.

Type of Sod Culture

The sod culture system practised in orchards may be of the following three methods:

(i) Sod Culture Method

In this method, grasses are allowed to grow on the orchard land in an undisturbed manner, *i.e.*, they are neither mowed down nor grazed by animals.

(ii) Sod Pasture Method

In this method, the orchard is treated both as a plantation of fruit crops as well as a pasture ground for grazing animals. Under Indian condition, this method appears to be more common where sod culture is practised.

(iii) Sod Mulch Method

This is a method of sod culture where the grasses grown in the orchard are mowed down but the cut out grasses are left on the land *in situ*.

(a) Advantages

(i) Like cover cropping, soil erosion is greatly checked in this method as the roots of grasses bind the soil in a more compact manner.

(ii) As movement of water over grassy area is much slower than bare land, sod culture is very effective in reducing surface run–off of water after heavy rain. Accordingly, this provides greater opportunity for downward percolation of water in the orchard soil.

(iii) The grasses hold sufficient amount of water between their blades (= laminae) during rainfall and later, this retained water seeps underground slowly. Sufficient amount of *dew* water is also retained by the laef–blades of grasses.

(iv) Where winter temperature goes low, the sodded area protects the land from getting too cold and similarly, when summer temperature goes high, the soil surface is protected from getting too hot.

(v) In sod mulch method, the mowed grasses left out on the orchard land provide a suitable mulch and adequate amount of *bio–mass* is thereby, added due to that.

(vi) Sod pasture method provides opportunity for grazing of animals. Adequate amount of excreta is simultaneously received by the soil when the animals graze. There is also a trend in some parts of India to maintain sod in the orchards for the purpose of penning cattle, sheep and goats as considerable amount of excreta is received from these animals in short time which serves as a good manure to the orchard land. It is estimated that a flock of 7000 sheep for a night may add dung equivalent to 15 tonnes of cattle dung. The dung dries up quickly for which loss of nitrogen due to ammonification is also checked.

(b) Disadvantages

Sod culture system of orchard soil management has demerits in most parts of India and the notable ones have been stated in the following:

(i) Hayes, 1970 in his classic book, *Fruit Growing in* India has quoted, "The effects of sod are imperfectly understood, but seem to be mainly a reduction is soil moisture and available nitrogen". According to this authority of Indian fruit culture, grasses and other vegetation draw much water from the soil and give it off into the atmosphere. Based on actual measurement, he opined that amount of moisture in the soil under sod remains much lesser than when clean culture is practised in the same soil. Another serious drawback what he stated is, amount of available nitrogen in the orchard soil is also decreased in sod culture. Thus, he did not support sod culture in plain lands of India but quoted that, "There is a great advantage in having sod in *hilly country* as it prevents erosion".

(ii) The shallow rooted fruit trees and shrub in particular, seriously suffer in sod culture due to the severe competition faced by their roots with those of grasses.

(iii) Quick death of some fruit trees by resorting to sod culture with Bermuda grass has even been on record under Indian condition.

Chapter 17

Management by Green Manuring

This is a system of orchard soil management which is done to add *bio–mass* (= organic matter) and *nitrogen* to the soil and thereby, aims at improving the textural condition as well as fertility of the soil. There are in fact, several methods of green manuring and these have been stated in the following:

(i) Seeds of leguminous plants belonging to one species are sown in the orchard land and when the seedlings emerged from them have grown to 40 to 50 cm height (depending on plant species) and are in succulent condition, the entire plants are buried into the soil by ploughing them down deeply or by spading however, in presence of adequate moisture so that they get rotten, *i.e.*, decomposed.

(ii) The practice remains the same as in (i), but seeds of more than one species of leguminous plants are sown.

(iii) This practice remains the same as in (i) or (ii) but seeds of non–leguminous plants are sown.

(iv) The seedlings are cut off with sickle or cutlass or, harvesting knife and then they are laid flat on the ground as such to get them rotten however, without ploughing over them.

Among the above, the *three conventional methods* of green manuring that are commonly practised have been described below in short.

Method 1

In this method, seeds of leguminous plants are sown and when the resulting seedlings are in very young stage, they are ploughed down into the soil under optimum moisture to ensure their rotting (= decomposition). The steps may be stated as follows:

(i) Selection of the Plant Species

The plant species which only belong to the botanical family, leguminoceae are used in this method. Even in this plant family, only some selected species, *viz.*, *dhaincha* (*Sesbania aculeata*), sunnhemp or *sann (Crotolaria juncea)*, cowpea (*Vigna catjong; V. sinensis*), cluster bean or *guar* (*Cyamopsis tetragonoloba*), horse gram or *kulthi* (*Dolichos biflorus*), moth bean or moth (*Phaseolus aconitifolius*), black gram or *urad* or *kalai* (*Phaseolus mungo*), green–gram or *mung (Phaseolus aureus)*, lentil or *masur (Lens esculenta; L. culinaris) etc.* and also some wild species which are quick growing and serve as more suitable materials as green manuring plants are given preference.

The selection of species is no doubt, an important factor. This is because, some species are known to have somewhat greater suitability in some regions. For example, cluster been and also *berseem* (*Trifolium alexandrium*) have greater adaptability in the northern and central parts of India, cowpea in Karnataka and Tamil Nadu, green gram and cowpea in eastern India, lentil in the Kashmir valley and perhaps in the Punjab and Haryana. Some species, notably, sunnhemp have however, wider adaptability and hence, This may be grown in almost all parts of India.

Apart from agro-climatic suitability, enriching fertility of the soil also depends on species. For example, black gram, moth bean, sunnhemp *etc.* add higher amount of nitrogen to the soil which may be more than 0.75 per cent, while cluster–bean, horse gram *etc.* add not more 0.35 per cent nitrogen. Besides, nitrogen, enriching the soil with phosphorous and potassium content is also relatively high when the land is green manured with cowpea, green gram, black–gram, sunnhemp *etc.*

Many farmers and fruit growers prefer to use more than one species of leguminous plants for green manuring. This may be a good practice provided that the rate of growth of the plants and their maturity to attain hardening of stems, *i.e., lignification* remain more or less the same and do not vary widely.

(ii) Season of Green Manuring

Green manuring can be done in any season of the year. But for the purpose of rotting, *i.e.*, decomposition of the plants when they are ploughed down, adequate *moisture* must be present in the soil. Therefore, rainy season is by far, the best time for the purpose. Besides, rate of growth of the plants is also maximum in the rainy season. The practice is in fact, most suitable when the seeds of the green manuring plants are sown just before the rainy season. In other words, sowing the seeds with the pre–monsoon shower is beyond doubt, most conducive. However, green manuring can be done at other season also if artificial irrigation is possible to be provided for decomposition of the plants. But winter does not appear to be suitable except in highly tropical situation as germination of the seeds and growth of the seedlings of the green manuring plants are hampered due to low temperature.

(iii) Preparation of the Land

Bringing the land to a suitable tilth by ploughing it at least once or preferably twice before sowing seeds of the green manuring plants is advisable to ensure quick and unhampered growth of the resulting seedlings rather than sowing the seeds without preparing the land.

Application of quick acting nitrogenous fertilizer like urea and if possible, little quantity of phosphatic and potash fertilizers along with that to the land while ploughing it is also advised by many to encourage quick and vigorous growth of the seedlings.

(iv) Sowing Seeds and Post Sowing Care

Only healthy, bold, fully mature and viable seeds of the green manuring plants should be used. To sow one hectare area of land with *dhaincha* seeds, about 35 kg of such good seeds may be required. The germination will be quicker if the seeds are soaked in water for few hours just before sowing. Sowing should be done by uniform broadcasting of them all over the land to be green manured.

The seeds germinate within a week and if any large patch is noticeable having no seedlings after a fortnight of sowing, it should be spaded or hoed and then sown with fresh seeds. The weeds, if are found to grow in large number during the first fortnight after sowing also need to be eradicated by pulling up with root and branch.

(v) Ploughing Down of the Plants

After about 1 ½ to 2 months of sowing when the resulting seedlings become 40 to 50 cm tall, they are *ploughed down* to the land. One ploughing is usually not enough to bury the seedlings thoroughly and hence, two or even three ploughings and rarely more than that may be required to do so.

It should be noted that stage of growth of the seedlings is a very important factor for ploughing them down into the land. The purpose of ploughing down or burying is to get them rotten, *i.e.*, decomposed. Accordingly, if the seedlings make more growth, their stems become *hardened, i.e.*, lignified and hence, rotting of them does not occur quickly. Thus, younger the seedling, more and more they are succulent and rotting of them takes place more quickly. Therefore, at earlier stage of growth, decomposition is quicker.

However, this method of green manuring is done not only to add organic matter but to enrich the soil inhabiting with addition of nitrogen also. This nitrogen addition is done by some soil in habiting bacteria which take up the gaseous, *i.e.*, free nitrogen present in the air between the soil practice. These bacteria feed on the roots of the leguminous plants, remain adherent there and ultimately die there. Due to such attachment or clinging on the roots, outgrowths known as *nodules* are formed to appear on the roots. When the seedlings are buried down or killed or cut off, these nodules which get enriched with nitrogen remain present under the ground and the soil is thereby, enriched with nitrogen.

Now, when the seedlings are in very young stage of growth, they decompose very easily and quickly as they are highly succulent. But the extent of nodule formation in their roots is lower and hence, ploughing down at this stage does not make adequate nitrogen addition to the soil. Therefore, ploughing over the plants is recommended when they are 40 to 50 cm in height, which is attained after 45 to 60 days of sowing. At this stage of growth, the extent of nodulation in their roots is optimum as well as they remain in succulent condition for which rotting takes place easily.

A very important aspect should be remembered in this connection, that is, adequate water must be present in the soil when the seedlings are ploughed down. This is because, if ample water is not available, the plants will not undergo rotting easily is short time and thus, the full benefit of green manuring is not obtained. The availability of adequate soil moisture exists if the ploughing down of the seedlings is done in full rainy season and if the downpour is satisfactory. But if done in other season, *i.e.*, not in rainy season, the benefit of rain is not obtained. Even in the rainy season also, if there exists a rainless period or of scanty rain when the seedlings have attained the required growth, their rotting may be difficult if ploughed down. In such cases, the only alternative is to irrigate the land by artificial means to ensure retention of optimum water in the soil so that rotting of the seedlings takes place easily. In order that the rain water or the irrigational water is collected and stored on the ground for rotting to take place, it is advisable that bundings with soil should to made on all sides of the part of land in the orchard where green manuring is to be done. By making such bunds in the inter-spaces, the fruit trees may also be saved from over-irrigation or water logging.

Method 2

This method is exactly the same as above in (1), except that the seedlings are cut back with sickle, harvesting knife, cut–lass *etc.* when they attain 40 to 60 days of growth. The cut out parts are left as such on the land without burying them by ploughing.

The practice of burying the seedlings by ploughing them as practised in the former method is not done here because of the fact that adequate moisture does not remain present in the soil at the time when the seedlings attain the right stage for burying them. In the former method, water is supplied in the absence of rain by artificial irrigation for the purpose of rotting of the seedlings. But in this method, if there is no rain, artificial irrigation is not provided due to difficulty or cost factor and the seedlings are only cut back. The cutting of the seedlings is done in order that they do not grow further and thereby, are not hardened. The decomposition of the seedlings takes place naturally but slowly on taking up the moisture that is available in the soil. However, it becomes faster if rain occurs after they are cut back. The nitrogen enriched nodules in the roots of the seedlings remain present in this method as the roots are not disturbed.

Method 3

In this method, leaves and succulent twigs of herbs, shrub or trees which are able to decompose easily are laid over the land and then buried by ploughing over them. Therefore, unlike leguminous plants, addition of nitrogen to the soil from the roots is not possible in this method, although addition of *bio-mass, i.e.*, organic matter is done due to decomposition of the leaves and twigs which have been buried.

The leaves and twigs which are to be buried should be chosen with care that they are not hardened (= lignified) and as far as possible remain succulent. Any plant may be used for this purpose and some examples may be cited as the top parts of *Gliricidia maculata, Pongamia pinnata, Erythrina indica, Ricinus communis, Ipomea carnea, Calotropis gigantia, Azadirachta indica, Polyalthia longifolia, Madhuca indica, Eichhornia crassipes etc.*

(a) Advantages

Green manuring done by growing and burying down leguminous plants in succulent condition is perhaps the *cheapest way* to add organic matter and nitrogen to the soil. By green manuring in this way, 15 to 25 tonnes of organic matter and 60 to 90 kg of nitrogen are possible to be added to the land in a hectare area.

Burying down succulent leaves and twigs also add adequate organic matter though nitrogen addition through root nodules is not possible.

It is an easy, cheap and highly effective method to *convert a clayee or a sandy soil* to a desirable and soft texture with adequate humus by green manuring done consecutively for several years.

When superphosphate is required to be supplied to the soil, it is better to apply it during green manuring. By this, the *clots of the fertilizer are broken down* and availability of phosphate to the crop grown on the land after green manuring becomes easy and takes place quickly.

(b) Disadvantages

There is hardly any disadvantage of green manuring. Nevertheless, weed growth sometimes becomes higher in a green manured land, especially if it is kept fallow, *i.e.*, no crop is grown there. In few cases, soil pests and diseases may increase due to green manuring. The benefit of green manuring is however, much more than these minor drawbacks.

Chapter 18

Management by Soil Coverage (= Mulching)

Mulching is a system of orchard soil management where the soil surface is kept covered with some material(s), which is done with the idea to reduce evaporation of water from the soil surface, to control growth of weeds, to protect the soil against extremes of heat and cold and partly to control erosion of soil. There are evidences, that mulching is one of the very old agricultural practices and had been practised shortly after organized agriculture was started about 12 thousand years back.

Quite a large number of materials are used for the purpose of mulching in different parts of the world. These materials can be broadly divided under two groups, *i.e., organic* and *inorganic*. The materials which are more commonly used under each of these two groups have been enlisted below.

Organic Mulches

The common organic mulches include compost, peat, fresh leaves and twigs of banana, sugarcane, mango, castor, many weeds and grasses like *sarkanda (Erianthus arundinaceus), bui (Aerva persica)* para grass *(Brachiaria mutico) etc.*, dried leaves and twigs of banana, sugarcane, water-hyacinth *etc.*, paddy and wheat straw, dried plants of maize, sorghum or other millets, saw dust, wood chopping, husks of groundnut, paddy *etc.*, newspapers or thick papers and the like.

Inorganic Mulches

Among the commonly used inorganic mulches, mention may be made to that of transparent, black or coloured polythene sheets, plastic film, reflective plastic film, resinous materials, petroleum spray or foam, stone chips, brick bats, brick dust *etc.*

Methods

The practice consists of covering only the surface of the land which is desired to be mulched with the material.

There is difference of opinion as regard thickness of mulch to be applied. Some orchardists are in favour of providing thick mulch to about 10 cm or even more becouse in such thickness, the effect of mulching is produced best. On the other hand, this is not supported by others for according to them, so thick mulching is not only unnecessary but also hampers the soil in getting aerated. The actual thickness depends much on the mulching material also.

Mulching may be done either permanently, *i.e.*, throughout the year or in some seasons only. Permanent mulching appears to have advantage in arid zone and seasonal mulching is usually done in other places. This is done especially during the winter and summer months to protect the soil against evaporation, extremes of temperature and high wind. It may be noted that the organic materials when used as permanent mulch are decomposed and hence, they become manure. In such case, it is referred to as *mulch manure* rather than as pure mulch. In seasonal, *i.e.*, temporary mulching, the mulching materials are taken out from the land when mulching of it is not desired.

For the purpose of mulching, either one type of material is used or two or more materials are combined together to derive the advantage of each. A combination of organic and inorganic mulches is also done. Of late, *bio–plastic* is also used for mulching. This is prepared by suitably blending maize flour with petrochemical polymer and has the property to destroy without leaving any residue with time or in contact with water. Hence, this materials may be used for short period mulching.

A system of mulching was formerly prevalent in pineapple plantations in the Hawaii island. About 100 to 200 metres long and 90 to 150 cm wide, asphalt impregnated heavy type of papers were used. Suitable holes were made on these papers at distances which corresponded to the planting distances. The papers were rolled over the soil surface leaving spaces between the strips. The holes on the papers represented the locations where planting were to be done and the propagating materials, *i.e.*, slips or suckers were planted in the soil through the holes to produce the plants. The papers remained there permanently as a mulch cover till the crop was harvested and then it was taken out or left to tear away. It was estimated that by this method, soil moisture conservation and weed control were done so effectively that pineapple might be grown when the rainfall was as low as 50 cm. the crop yield had also been estimated to increase by more than 20 per cent, besides that production of slips was higher. The method is however, rarely practised now for many drawbacks.

If polythene sheet is used as a mulch cover, it is better to use that of 200 gauge than the 100 gauge as the former does not tear away easily and has high durability. Before wrapping the soil with such a sheet of mulch, it is essential that the soil should be pulverized and levelled, because the polythene may tear away if clods are present on the soil surface. In order that the polythene sheet is not displaced from its position due to wind after it is wrapped, it is advisable that shallow trench should be dug in the soil on all sides at the boundary of the polythene sheet which has been wrapped. On doing so, the borders of the sheet are suitably placed inside the trenches and then the trenches are thoroughly filled with soil. By this type of embedding the borders of the sheet it will remain in its position and will not be blown away by wind.

For other mulching materials, only placement on the soil surface serves the purpose but putting some bricks or stones here and there should be done to retain the mulches in proper places without being blown away. This should be done especially when paper or cardboard mulch is used.

Mulching may be done either all over the soil surface in a continuous manner, *i.e.*, without leaving any gap, or may be done leaving suitable gaps. Leaving some gaps between mulch covers appears to be beneficial for aeration of the soil and is more commonly practised. On the other hand, continuous mulching without leaving gaps seems to be advantageous during the summer season in arid zone and winter season in temperate zone to ensure thorough protection, respectively against high temperature and frost or snow.

Lastly, it also needs pointing out that in mulching technology, personal experience of the orchardist is an important factor to judge what mulching material is to be used, its thickness, duration of application, whether it is to be applied leaving gaps or not and other aspects, in order to derive the maximum benefit.

(a) Advantages

A number of benefits of mulching are derived in fruit culture, among which the following are mentionable:

(i) A suitable mulch cover is undoubtedly an effective means to reduce evaporational loss of moisture from the soil surface. Accordingly, the practise is of great help especially for the shallow-rooted fruit crops like papaya, litchi, sapota, avacoda, *phalsa etc.* The roots of these fruit trees or shrubs which are concerned with uptake of water and nutrients in solution form from the soil and are known as feeder roots, abound very close to the soil surface and they often suffer from moisture stress whenever there is little unavailability of water in the top soil as a result of evaporation. A mulch cover on the soil surface effectively protects these fruit crops against moisture stress by reducing the evaporational loss from the soil surface.

(ii) Mulching prevents or reduces the emergence or growth of weeds to a considerable extent, depending on the type of mulch, the thickness of mulching and other factors.

(iii) When manures, peat or other humus forming organic materials are used as mulch, they add nutrients to the soil, loosen the soil surface and gives a sponge action such that during rainfall or irrigation, they absorb sufficient quantity of water and later permit it to seep into the soil slowly. With such mulching materials, surface run off of water is also reduced.

(iv) In a fruit plant like strawberry, whose fruits are very soft and lie over the ground (as the plants grow by producing runners), mulching greatly helps to remain them clean and protect against rotting. *Straw mulching* is a common practice in culture of this fruit crop and hence, the name *strawberry*. In other creeping fruit plants also like water melon and musk melon, mulching is found much helpful for fruits of better quality.

(v) Mulching is a highly effective device to protect the soil against very high temperature in summer and very low temperature and frost or snow in winter. For strawberry and other temperate fruit crops, it therefore, renders great benefit.

(vi) The growth of roots has been found to be increased in many fruit crops by mulching.

(vii) Increase in size of fruits and decrease in cracking abnormality of them have been reported in litchi by mulching the bases of the trees (= basins). This effect is produced as litchi being a shallow rooted fruit tree, its feeder roots are protected from moisture stress due to the presence of the mulch cover.

(viii) It has ben observed in Hawaii that in pineapple plantations, nitrification in the soil is much increased by the use of paper mulch.

(ix) The experiments conducted in Himachal Pradesh have provided evidence that mulching with polythene and pine needles at the rate of 10 tonnes per hectare in strawberry field had increased yield of the fruits by 68 and 33 per cent respectively over control and had also included early ripening.

(x) Mulching with black coloured polythene is known to offer many advantages. The effect of this mulch in retarding loss of soil moisture, weed growth in the gaps between mulch covers, improvement of soil structure, lowering of soil temperature in day time but not much lowering of temperature at are night greatly pronounced in most cases.

(xi) Some mulches like that of *reflective plastic* decrease soil temperature to a considerable extent by reflecting light and heat. Mulching done with petroleum spray or resinous material also produces similar effect.

(xii) A soft mulch cover protects the soil surface against crust formation which is caused due to open exposure of it.

(xiii) The soil structure is also improved by a mulch by reducing the raindrop effect.

(xiv) Some of the mulches also protect the soil against erosional loss, especially erosion due to wind. Polythene mulch is very effective in this respect in preventing blowing away of the surface soil by wind.

(xv) The organic materials when mulched permanently on soil surface ultimately decompose and thereby, turn into *organic manure*. Thus, in course of time, these mulches improve textural condition and fertility of the soil by serving as manure.

(b) Disadvantages

Despite many advantages, mulching has some drawbacks also and the notable ones have been stated below:

(i) Mulching sometimes create a breeding centre for many insects and other invertebrate animals. Some of the soil borne pathogenic fungi and bacteria also multiply quickly when mulching is done. In fact, the warmth and humidity that are maintained below the mulch cover become ideal for survival and growth of these insects, animals and disease organisms. This adverse effect is greater when organic mulches are used. This is because, the organic mulches provide food and moisture to these insects, animals or micro-organisms and hence, they multiply greatly at quicker time. Application of plant protection chemicals occasionally over these mulches is therefore, advisable to protect against these pests and diseases.

(ii) Surface irrigation by flooding, check or bed, furrow, border, basin and ring has difficulty in the orchard where mulching is done. For drip irrigation also, special measure is to be taken if the land has been mulched.

(iii) Aeration of the soil may be hampered if mulching is done in an improper manner.

(iv) Intercultural operations and harvesting of fruits from the orchard trees have difficulty where mulching has been done. Some mulches may create difficulty in walking on the orchard land as well.

(v) Inorganic mulches, especially if done with polythene sheet and other plastic materials which are highly inflammable have a great drawback in that, they catch fire easily. It may be noted that throwing a glowing match stick or a cigarette end over such a mulch cover may even produce a conflagration in the orchard. Adequate precaution is therefore, necessary if such type of mulches are used.

Chapter 19

Sanitary Measures

The orchardist should look after the floor of his orchard as the floor of his own house. He should not forget the usage, "Prevention is better than cure" and sanitation of the floor keeps off many undesirabilities in his orchard. Orchard sanitation should in fact, be maintained at every footstep from the start of the orchard and till it lasts at that site. Among other considerations, the following in particular, should be given high emphasis.

1. At the entrance gate of the orchard, *lime dust* should occasionally be sprinkled so that everyone who goes inside it tramps it while entering.
2. Before planting a propagating material, the orchardist should be sanguine of the fact that it is not a diseased material and is free from the attack of insects or other pests. Even a broken material should be discouraged as there may be *opportunistic infection of disease organisms* through the damaged parts.
3. After pruning operation of the fruit trees or shrubs, the pruned out branches and leaves should not be left on the orchard soil. This is because, if these parts contain disease organisms, the healthy plants may be infected. In grape vine, the anthracnose disease (popularly known as *karpa*) which is caused by the fungal species, *Gloeosporium ampelophagnum*, and in guava caused by *Gloeosporium psidii* fungi very often attack healthy plants through pruned parts infected with this disease.

4. Stagnation of water in any part of the orchard even if it is shallow should always be discouraged. To avoid this, irrigation and drainage channels in the orchard should be checked regularly to ensure that water flows through them freely and is not blocked anywhere. If water logging results due to any depression of soil, it should be filled with soil so as to bring it to the ground level.
5. Damaged fruits harbour insect pests and diseases. Therefore, such fruits which have been partly eaten by birds or have fallen from the trees if noticed, should at once be removed out from the orchard. (Prof. W.B. Hayes*[1]
6. If any branch of any tree is found to be broken, it should be cut back (= headed back) and the cut surface of the intact branch should forthwith be smeared with fungicidal paste. This is because, trees are very often infected by disease causing pathogens through broken branches.
7. The orchardist or the employees should everyday make a round all round the orchard. If any undesirable material such as a mummified guava fruit (guava fruits become mummified, *i.e.*, turn black and very hard due to anthracnose disease) or even a dead bird come to the sight, it should be taken out of the orchard.
8. If rat holes are found in the orchard soil, effective rhodenticides like zinc, magnesium or calcium phosphide, warfarin, bromadiolome, coumatetralyl, chlorfacinone *etc.* in tablet or liquid form should be poured in the holes and plugged with soil.
9. Orchard floor should never be used to store any material such as manure, fertilizers, bamboos, bricks, stone chips *etc.* even for a short period also.
10. It is advisable to provide a lavatory at any corner in the orchard for use of the employees especially if the orchard area is large.

1 Prof. W.B. Hayes, the American Missionary was associated with the Horticulture Department of the erstwhile Allahabad Agricultural Institute which is now transformed to the Sam Higginbottom University of Agriculture and Technology. He brought from the U.S.A. many varieties of grapefruit, sweet orange, mandarin orange, tangelo, tangor, lemon, *etc.* and planted those in the orchard of the Allahabad Agricultural Institute in the 1940s. Indian soil had thus, the opportunity for the first time to nurture these *Citrus* species and hybrids. Prof. Hayes donated all his possessions for improvement of the Allahabad Agricultural Institute. He spoke his last words on this Earth to his students while teaching them at the same Institute.) While taking practical class of the students at the Allahabad Agricultural Institute he had the chance to notice that one student plucked a guava fruit from a tree, ate it partly and then threw it away on the orchard soil. Prof. Hayes told him that he could pluck as many guavas as possible and eat there but he could not have any right to throw away any half eaten guava on the orchard floor because such an act might only infect the healthy trees with diseases. The boy felt ashamed and assured him non–recurrence of such an act in his life).

Appendix I

Botanical Names of Plants Referred to in the Book

Aonla: *Phyllanthus emblica* L.
Apple: *Malus domestica* Mill
Apricot: *Prunus armeniaca* L.
Banana: *Musa* species
Bael: *Aegle marmelos* Correa
Barbados cherry: *Malpighia emarginata* DC
Ber: *Ziziphus mauritiana* Lam.
Bermuda grass: *Cynoden dactylon* (L.) Pers.
Betel vine: *Piper betle* L.
Bullock's heart: *Annona reticulata* L.
Calamondin: *Citrus madurensis* Lour.
Cape-gooseberry: *Physalis peruviana* L.
Cashew: *Anacardium occidentale* L.
Castor: *Ricinus communis* L.
Cherry: *Prunus avium* L.
Citruses: *Citrus* species
Custard apple: *Annona reticulata* L.

Date: *Phoenix dactylifera* L.

Dragon fruit: *Selenicereus undatus* (Haw.) D.R. Hunt

Eucalyptus: *Eucalyptus globules* Lab.

Gladiolus: *Gladiolus* species

Grape: *Vitis vinifera* L.

Grapefruit: *Citrus paradisi* Macf.

Guava: *Psidium guajava* L.

Jackfruit: *Artocarpus heterophyllus* Lam.

Jamun: *Syzygium cumini* Skeels

Karonda: *Carissa carandas* L.

Kiwi: *Actinidia deliciosa* (A. Ch.) C.F. Li. & A.R. Fer.

Lemon: *Citrus medica* var, *limonum*

Litchi: Litchi chinensis Son

Longan: *Dimocarpus logan* Lour.

Maize: *Zea mays* L.

Mandarin orange: *Citrus reticulata* Blanco

Mango: *Mangifera indica* L.

Mangosteen: *Garcinia mangostana* L.

Musk melon: *Cucumis melo* L.

Papaya: *Carica papaya* L.

Pear: *Pyrus communis* L.

Phalsa: *Grewia subinaequalis* DC.

Pineapple: *Ananas comosus* Merr.

Pomegranate: *Punica granatum* L.

Potato: *Solanum tuberosum* L.

Pummelo: *Citurs grandis* Osbeck

Redwood: *Sequoia sempervirens* (D. Don) End.

Sapota: *Manilkara zapota* (L.) P. Roy

Sorghum: *Sorghum bicolor* (L.) Moe.

Sour lime: *Citrus aurantifolia* Swingle

Sweet lime: *Citrus limettioides* Tanaka

Strawberry: *Fragaria ananassa, Fragaria* species and Hybrids

Sugarcane: *Saccharum officinarum* L.

Sweet orange: *Citrus sinensis* Osbeck

Water hyacinth: *Eichhornia crassipes* Mart.

Water melon: *Citrullus lanatus* (Thumb.) Mat. & Nak.

West Indian cherry: *Malpighia emarginata* DC

Appendix II

Irrigational Particulars

(A) Lifting of Water (Approximate value)

Water Lifting Device	*Rate of Discharge (litre/hour)*	*Depth of Lifting (metre)*
Egyptian screw	22730.00	1.00
Persian wheel	18184.00	10.70
Sewing basket	18184.00	1.00
Suction pump	40914.00	7.00
Centrifugal pump (5 cm)	22730.00	7.00
Centrifugal pump (7.5 cm)	45460.00	8.00
Centrifugal pump (10 cm)	86374.00	9.00

(B) Duty of Water

Formula to determine duty of water:

$$D = \frac{1 \times 60 \times 60 \times 24 \times (b)}{(V)}$$

where,

D = Duty of water in acre of crop matured by 1 cu–sec of water discharged without break for a given period of time.

(b) = Number of days during which 1 cu–sec of water flows to mature the crop.

(v) = Volume of water (c-ft) to mature crop in 1 acre.

(C) Conversion of Units

1 acre-inch of water	=	22687.5 gallons
	=	103137.37 litres
1 cu-sec of water	=	6.25 gallons per second
	=	28.41 litres per second
1 c-ft of water	=	6.25 gallons
	=	28.41 litres
	=	28.350 kg.

Appendix III

Conversion Factors

Length

1 kilometre	=	10,000 metres = 0.621 mile = 1094 yards = 3281 feet.
1 metre	=	0.001 kilometre = 10 decimetres = 100 centimetres
	=	1000 millimetres = 1.0936 yards = 3.281 feet = 39.37 inches.
1 mile	=	8 furlong = 1760 yards = 5280 feet = 1.609 kilometres.
	=	1609 metres.
1 yard	=	0.9144 metre = 9.144 decimetres = 91.44 centimetres
	=	914.4 millimetres = 3 feet = 36 inches = 2 hands (approx.)
	=	4 *bighas* (approx).
1 foot	=	0.3048 metre = 3.048 decimetres = 30.48 centimetres
	=	0.33 yards = 12 inches.
1 inch	=	0.0254 metre = 0.254 decimetre = 2.54 centimetres
	=	0.083 foot.

Area

1 hectare	=	100 areas = 10,000 square metres = 2.4711 acres
	=	0.8237 *bighas* = 0.4114 *cottahs.*
1 square mile	=	640 acres = 259 hectare.
1 square metre	=	10.765 square feet = 1550 square inches.

1 acre	=	0.4047 hectare = 4840 square yards = 43,560 square feet
	=	3 *bighas* (approx.) = 60 *cottahs* (approx).
1 square foot	=	0.0929 square metres = 144 square inches.
1 square inch	=	6.452 square centimetres.

Index

E

F

G

H

I

L

M

N

O

P

R

S

T

U

V

W

X

9 789359 199740

Printed at Repro India Ltd.